The Rentier State in Africa:

Oil Rent Dependency & Neocolonialism in the Republic of Gabon

Douglas A. Yates

Africa World Press, Inc.

P.O. Box 1892
Trenton, NJ 08607

P.O. Box 48
Asmara, ERITREA

Africa World Press, Inc.

P.O. Box 1892
Trenton, NJ 08607

P.O. Box 48
Asmara, ERITREA

HD
9577
G22
Y38
1996

Book design: Jonathan Gullery
Cover design: Linda Nickens

Library of Congress Cataloging-in-Publication Data

Yates, Douglas Andrew.
    The rentier state in Africa : oil rent dependency and
neocolonialism in the Republic of Gabon / Douglas Andrew Yates.
      p. cm.
    Includes bibliographical references and index.
    ISBN 0-86543-520-0 (cloth). - - ISBN 0-86543-521-9 (paper)
    1. Petroleum industry and trade - - Gabon. 2. Gabon - - Economic
conditions - - 1960- 3. Gabon - - Colonial influence. 4. Gabon-
-Dependency on foreign countries. I. Title.
HD9577.G22Y38 1996
330.96721'04–dc20   96-14980
                                                CIP

CONTENTS

Acknowledgements

Books are often collaborative efforts that benefit from the combined efforts of many people. I would like to thank my professors, Irene Gendzier and Edouard Bustin of Boston University; my publisher, Kassahun Checole, and my editor, Frank Blisard of Africa World Press; contributors, Martin Keys, Adeyeke Adebajo, Scott Osberg, Sarah Stiles, John Guardiano, Shaheen Mozaffar, David Gardinier, James Barnes, Guy Marcel Eboumy, Geoffrey Wallace, Tamara Scott, Rachael & Richard Coulston; my parents William & JoAnn, my brother James, and my wife Corentine.

This book is dedicated
to all those who have died
in the struggle for Democracy
in Gabon.

Introduction

The oil business, indispensible to the world's economy, is arguably the largest and most widely integrated of all the global commodity networks. Founded by industrial "robber barons" in the nineteenth century, oil corporations have grown to become the very symbols of big business in the twentieth. It is hard to overstate the importance of oil to the capitalist system: oil is an essential input into practically every industrial process; it is the blood of mechanized production, and the fuel of motorized societies; and when measured in terms of volume, it is the number-one commodity traded in the world. Oil is shipped to literally every country by a handful of corporations— the so-called "majors"—which continue to virtually monopolize its downstream operations.[1]

Many Africans are increasingly concerned with the role that these majors are playing in African countries. They ask themselves "What has the oil business done for Africa?" Some see big oil as a kind of blessing that has given its producer countries the cash they needed to accelerate economic development. But others see big oil as a curse that has retarded economic development and thwarted political change. Still others see big oil money itself as the real problem.

This is the story of how Africans deeply involved with

the international petroleum industry benefitted and suffered from this kind of involvement. It begins with exploratory offshore drilling by the majors in the 1950s. It covers the remarkable era of decolonization in the 1960s, when African leaders took control of their national oil reserves. It tells the story of how the oil booms of the 1970s and the oil price collapse of the 1980s shaped the political discourse on democracy and development. Finally it brings us to the 1990s, as these "*rentier*" states face the prospects of resource depletion and a future without oil.

While this is primarily the story of the Republic of Gabon, it conveys important lessons concerning other oil states in Africa such as Nigeria, Angola, Congo, and Cameroun. The story of Gabonese oil therefore is intended to tell the story of African oil; and the story of the rentier state in Gabon, the story of the rentier state in Africa.

THE VALUES

Some Africanists feel that a disciplinary bias toward theoretically driven studies of Africa has "diverted scholarly attention away from the burning issues of social and imperial domination and toward far less controversial subjects."[2] In a recent article in *Issue*, Richard Sklar suggested that "[s]cholars do not need theories or methodologies to help them identify the problems of society. Principles, or values, will suffice" (p. 20). His point is that, "[i]n the human sciences, moral sentiments determine the problems that we select for resolution and the relative importance that we attach to them" (p. 21). Theory without morality is an empty thing.

What are the values that motivate the present study? That make it meaningful and give it relevance to Africa and Africans? The first is *democracy*, by which is meant a significant participation of ordinary people in the decision-making processes of the state. Democracy in Africa may be an ideal rather than a reality, but it is nevertheless an ideal by which ordinary Africans can evaluate the political actions of their elites. It is also an ideal that can provide them with a vision of a better tomorrow. The second value is the principle of *economic justice*, which in the

African context means, among other things, a more equitable distribution of wealth. Nowhere is the gap between rich and poor greater than it is in Africa. Since there can be no political democracy without economic democracy, some kind of significant social and economic redistribution will be necessary in these African countries before any kind of meaningful democratic change will occur. The third value is *national self-determination*, which means the end of foreign domination of African economies and governments, and the beginning of a new kind of freedom. Genuine sovereignty will allow Africans to decide for themselves what they want from life, and to act in our international system as the independent states they are purported to be.

THEORIES OF DEVELOPMENT

There are several critical schools of thought that not only provide theoretical explanations for the problems associated with oil dependendcy, but also have played historically significant roles in the thinking about African development. At the risk of boring the reader,[3] I will briefly summarize them here in order to explain where rentier theory fits in to the existing critical literature on development.

One such school is the "*dependency and underdevelopment*," associated with André Gunder Frank, Theotino Dos Santos, and Samir Amin.[4] These men assume the existence of a capitalist "world-system" that is conceptually divided into two concentric spheres: an advanced capitalist core and an underdeveloped periphery. The relationship between the core and the periphery serves to impoverish the former and enrich the latter, so that the peripheral countries have been "*underdeveloped*" (in the active sense) by the same core states upon which they depended.

In this model, the *peripheral* (oil-exporting) states have been exploited by the *core* (oil-consuming) states upon which they have depended. In this school, depending upon whom one reads, the agents of exploitation have been variously identified as states, banks, or multinationals.

However the result was the same: Underdevelopment in the oil-producing states is caused by their condition of dependency.

An alternative theory, termed *"dependent development,"* is associated with Fernando Cardoso, Peter Evans, and Thomas Bierstecker,[5] who challenge the notion that dependency necessarily produces underdevelopment. Otherwise, they ask, how does one account for the success of any of the newly industrialized countries that emerged in the so-called periphery of the world-system? Dependent developmentalists agree that multinational corporations and other agents of Western capital have created conditions of "dependency" in the Third World, but — rather than argue that multinationals have thereby created some kind of underdevelopment — the dependent developmentalists argue that capital formation and economic development were actually stimulated by the process. These scholars have focused on big states in the semi-periphery, states such as Brazil and Nigeria that have achieved *development under conditions of dependency* — or, in other words, "dependent development." The growth of high-technology oil sectors (which has enabled states in these countries to expand their control over the key sectors of their economies) is an example of dependent development, not underdevelopment, in this model. An important relationship has then emerged between the state, local capitalists, and transnational corporations which Peter Evans terms the "triple alliance." The real problem, as seen by these authors, is how the advances made possible by a "triple alliance" can be generalized, consolidated, and improved.

"Neomarxists" suspect the dependency theorists of advocating a nationalist mythology that is both misleading and dangerous. Neomarxists such as Bill Warren, Anthony Brewer, and Colin Leys, returned to the classical Marxist notion that *capitalism itself is development.*[6] They object to the implicit claim of dependency theorists that development must occur only under conditions of national autarky. On the contrary, capitalism did not have to be an instrument of nationalism to be development. The real problem, according to this line of reasoning, is not that gen-

uine capitalist development has failed to occur in the Third World, but rather, that it *has* occured.

Each of these developmental theories has its strengths, but each also has difficulties explaining the particularities of oil-producing countries.

"Dependency-and-underdevelopment" theory, for example, failed to adequately describe the oil-rich states. The wealth of Kuwait, for example, is hard to explain with a simple mechanical model of the world-system. How did a peripheral country like Kuwait acquire so much wealth from the core? Peripheral accumulation violates the core-centered direction of world capital flows. Does accumulation sometimes happen in the periphery? The oil-states have to be treated as exceptions to the rule, hence the rule does not adequately explain these cases.

"Dependent development" theory also fails to describe any of the less populous oil-rich states (i.e. low-absorbers). Qatar, for example, may be cash-rich, but is by no means "developed" in any serious sense of the word. Small oil emirates can not experience the dependent development of their more populous neighbors. Nor could a Gabon ever aspire to be a Nigeria.

None of these theories, furthermore, account for the *neopatrimonialist* realities prevelant in Africa.[7] Jean-François Médard has observed that in Africa "public authority has been made an object of appropriation by the formal officeholders, functionaries, politicians and military personnel, who based their strategies of individual ascendancy or family ascendancy on a private usage of the res publica."[8] Such personal observations delve into the historical specificity of the African states, and thus get at the *foundations* of the dependency debate.

"At the end of the seventies," writes Médard, "an increasing dissatisfaction arose among a number of French Africanists, mostly political scientists, opposing the prevalent approach to African studies, the dependency approach."[9] Scholars such a Jean-François Bayart have attempted to escape from the projection of external determinism and the negation of historicity of African societies by venturing into the dynamics of African societies.[10] But should we abandon the search for generalizeable development theories in international relations when we discuss Africa?

The *"theory of the rentier state"* is useful to scholars who are interested in patterns and problems of development specific to petroleum-rent-dependent states, not only in Africa, but throughout the developing world. Rentier theory posits that the conditioning factor of economic stagnation and political authoritarianism in oil-dependent states is the corrosive effect of external rent. Rentier theory is as concerned with internal developments (such as the emergence of a rentier mentality and rentier class) as with external. But what is interesting is the fact that rentier theory was developed by Middle Eastern scholars who were studying the effects of oil on the nature of Arab states. That a theory which emerged out of historical studies of the Middle East can be applied to historical realities of sub-Saharan Africa says something encouraging about the possibility of crossing the Saharan conceptual divide.

Those of us who have studied the international oil business have long suspected that cultural differences between Northern and sub-Saharan Africans did not substantially limit the globalization of petroleum development. The trend in world oil has been internationalization rather than regionalism. There is no need therefore to speak of African oil in a specifically "African" political-economy approach, because the continent has been integrated into the world oil economy by the majors in such a way that certain recognizable patterns will emerge despite historical specificity.

FRANCOPHONIE AND THE BEAST

France is one of the few world powers that continues to play an active role in postcolonial Africa. "French influence and involvement in the affairs of the continent," writes Daniel Bach "remain far more significant than those of other former colonial powers."[11] What accounts for the participation by France in contemporary African affairs?

One explanation is that African economies make good dumping grounds for surplus French goods. Related to this is the fact that African countries are also excellent sources of cheap raw materials. A French euphemism frequently used to describe/disguise this kind of crude self-interest is the word *"coopération,"* which suggests a voluntarism that may or may not in fact exist. This policy has, through such practices as regional trade blocs and the CFA franc, preserved former French African colonies as a profitable marketplace for the French business class, while providing the necessary illusion that it fosters "cooperative" development.

Another reason is that the political class in France entertains the notion that influence over African affairs places France at the summit of world affairs, equal to if not greater than other major world powers. The romantic euphemism for this is *"grandeur."* As Daniel Bourmaud has suggested, "the best way to account for French policy in Africa seems to be the simple search for power, with the objective of achieving 'great power' status in the international system."[12] Daniel Bach suggests, more modestly, that "Africa remains the only area of the world where France retains sufficient influence as to guarantee its claims to middle power status in the international system."[13]

This ambition proved as important for the French left of the 1980s as it is for the French right of the 1990s. The African policies of François Mitterrand over time became indistinguishable from those of his Gaullist predecessors, Giscard, Pompidou, and de Gaulle. Jean-François Bayart has discussed this in great detail in his study of the African policies of Mitterrand.[14] The fact that president Jacques Chirac has retained Jacques Foccart as his advisor on African affairs speaks to the continuity of French African policy. De Gaulle's vision of Africa, as Edouard Bustin suggests, has remained the policy preference of succeding regimes in France.[15]

Is the French political class justified in entertaining this notion? Who knows? Maybe "affairism" does give them the prestige they seek. Maybe France is recognized as an important player in the African arena. Maybe interventions demonstrate French military capacity. But is the recognition and self-esteem that the French gain worth the sovereignty that Africans lose?

In our world of ecodisasters, genocides, pestilences, plagues, interventions, and wars, perhaps it *is* hard to work up a good old-fashioned indignation at neocolonialism any more. Some would even go so far as to suggest that if African countries benefit from their dependent relations with France, then could dependency really be all that bad? If there is a penchant to romanticize the colonial past, then it will be very hard for Africanists to decry neoclonialism as an evil.

What this book will show, in no uncertain terms, is *how* and *why* neocolonialism causes underdevelopment, dictatorship, and general suffering for the average African who lives under it. While the patronage and protection of a neocolonial power does have its advantages (e.g., regime stability and financial aid), *the need to satisfy the interests of a neocolonial power places a heavy burden on its African subject society — in some cases destroying the very foundations of economic and political development.*

For those readers who are already familiar with France's activities in so-called "independent" African countries, there may be few factual surprises in the following pages. However, the malaise of our time, reflected in the postmodern relativism of much recent scholarship, is not the lack of critical information, but rather the inability to become genuinely outraged at injustices. One must first consider neocolonial structures of domination "unjust" beforeone can become outraged at them.

In the past, the role of the intellectual was, in princple, to "speak truth to power" — to act as a check on crude realism and as a voice of moral ideals. In this century, instead, many intellectuals have assumed the role of representing the ideas of imperial powers, "speaking power to truth." Realists in international relations contend that there have always been dominators and dominated. Realists contend that international relations is the search for power. If there were always oppressive men and oppressive deeds, then at least intellectuals assumed the role of speaking out against such men and such deeds, even if only in principle. These days it is fashionable in some quarters to justify them.

We might not be able to end neocolonialism by challenging it in books, but we might expect ourselves to be against it anyway. Julien Benda lamented in the early part of this century, when discussing the political responsibility of intellectuals, that: "Orpheus could not aspire to charm the wild beasts with his music until the end of time. However, one could have hoped that Orpheus himself would not become a wild beast."[16]

NOTES

1. The term "downstream" operations refers to such activities as refinment, transportation, and marketing of petroleum products, as opposed to "upstream" operations, which refers to the drilling and pumping of oil.
2. Richard L. Sklar, "The New Modernization," *Issue: A Journal of Opinion* 23/1 (1995):19
3. Not to imply that development theory is a boring subject. For an interesting discussion of theories of economic and political development see: Irene Gendzier, *Development Against Democracy: Manipulating Political Change in the Third World* (Washington D.C.: Tyrone Press, 1995).
4. For a good review of the ideas of these thinkers, see: André Gunder Frank, *Dependent Accumulation and Underdevelopment* (New York: Monthly Review Press, 1979); Theotino Dos Santos, "The Structure of Dependence," *American Economic Review 60*, no. 2 (May, 1970): pp. 231–236; Samir Amin, *Unequal Development: An Essay on the Social Formation of Peripheral Capitalism* (New York: Monthly Review Press, 1976).
5. Major definitive works in this school of thought include the following: Fernando Henrique Cardoso and Enzo Faletto, *Dependency and Development in Latin America* (Berkeley: University of California Press, 1978); Peter Evans, *Dependent Development: The Alliance of Multinational, State, and Local Capital in Brazil* (Princeton: Princeton University Press, 1979); Thomas J. Bierstecker, *Multinationals, the State, and Control of the Nigerian Economy* (Princeton: Princeton University Press, 1987).
6. For the groundbreaking work in this school, see: Bill Warren, "Imperialism and Capitalist Industrialization,"

New Left Review, no. 81 (September-October 1973). See also: Anthony Brewer, *Marxist Theories of Imperialism: A Critical Survey* (London: Routledge and Kegan Paul, 1980); Colin Leys, "Capital Accumulation, Class Formation and Dependency— the Significance of the Kenyan Case" in Ralph Miliband and John Saville, eds., *The Socialist Register* (London: Merlin Press, 1978), pp. 241–266.

7.   Jean-François Médard has defined neopatrimonialism as follows: "Patrimonialism being an ideal subtype of traditional domination corresponding to a lack of differentiation between what is private and what is public, neo-patrimonialism is a mixed type combining in various degrees differentiation and lack of differentiation of the public and the private domains." From "Politics From Above, Politics From Below," a paper presented at the NFU Conference: State and Locality (Oslo, June 18–20),: p. 18.

8.   Jean-François Médard, "Le Big Man en Afrique: Esquisse d'Analyse du Politicien Entrepreneur," *L'Année Sociologique* 42 (1992): 167.

9.   Jean-François Médard, "Politics From Above, Politics From Below," p. 10.

10.   Jean-François Bayart, *L'Etat en Afrique: La politique du ventre* (Paris: Fayart, 1989).

11.   Daniel Bach, "France's Involvement in Sub-Saharan Africa: A Necessary Condition to Middle Power Status in the International System," in A. Sesay, *Africa and Europe* (London: Croom Helm, 1986):, p. 75.

12.   Daniel Bourmaud, "France in Africa: African Politics and French Foreign Policy," *Issue: A Journal of Opinion* 23/2 (1995): 61.

13.   Bach, p. 75.

14.   Jean-François Bayart, *La politique africaine de François Mitterrand* (Paris: Karthala, 1984).

15.   Edouard Bustin, "Une Certaine idée de l'Afrique: De Gaulle's Vision of Africa between mythology and pragmatism," presented at Boston University, "Brazzaville +50: An International Conference on Francophone Africa and Franco-African Relations," Groupe de Recherches sur l'Afrique Francophone (Boston: 7–8 October, 1994).

16.   Julien Benda, *The Treason of the Intellectuals (La Trahison des Clercs)* (New York: Norton, 1969), p. 199.

Chapter I

The Theory of the Rentier State

INTRODUCTION

The theory of the rentier state is a complex of associated ideas concerning the patterns of development and the nature of states in economies dominated by external rent, particularly oil rent. The purpose of this chapter is to provide the essentials of the arguments, beginning with an historical summary providing background to the theory and its authors, then analyzing various component ideas, including definitions of such key terms as rent, rentier, rentier economy, rentier mentality, and rentier class.

THE RENTIER STATE

The concept of the "*rentier state*" was postulated by Hossein Mahdavy with respect to pre-revolutionary Pahlavi Iran in 1970; the idea has since been appropriated by a community of Middle East specialists in their discussion of the Arab world.[1] The theory in its broadest sense defines rentier states as those *countries that receive on a regular basis substantial amounts of external economic rent.*[2] It does not define the rentier state as exclusive to the Persian Gulf or the Middle East. But rentier theorists have had Arab oil-exporting and oil-transiting states in mind, particularly during the historical period 1951–1981, when they appropriated a larger share of the economic rents associated with their petroleum industries. In 1951 the god-father of populist petroleum politics, Mohammed Mossadegh, nationalized Anglo-Iranian and founded the National Iranian Oil Company—one model for the nation-alization of the Suez Canal by Nasser (1956), which in turn set a precedent for Northern and sub-Saharan African nationalizations such as Algeria (1967), Libya (1970), Nigeria (1970), and the focus of this work, Gabon (1974).

For Mahdavy the period 1951–1956 represented a watershed (or what he called a "landmark") in the eco-nomic history of the Middle East: an historical era when radical nationalizations transformed ex-colonial models of petroleum exploitation into what he termed "fortuitous etatism."[3] Massive amounts of foreign currency and credit generated by petroleum development flooded into the state coffers and, Mahdavy argues, turned at least some oil-producing countries into rentier states (e.g., Iran).

For Mahdavy, "the stage at which a country can be called a rentier state is determined arbitrarily," but he is primarily interested in cases in which "the effects of the oil sector are significant and yet the rest of the economy is not of secondary importance."[4] Kuwait and Qatar he cites as extreme examples of the phenomenon, with lim-ited capabilities for industrialization and few alternatives to rentierism. The case of Iran is important to Mahdavy precisely because, given its size and potential, it had alter-natives which the extreme cases lacked.

The idea of the rentier state assumed new importance in the decades following Mahdavy's work. Mahdavy was an economist writing with an eye on the events of the 1950s and 1960s who could not have forseen the dramatic trans-formations which were about to befall the world oil indus-try in the 1970s. The growing strength of the OPEC cartel, the Arab oil embargo of 1973, and the fall of the Shah of Iran in 1979 put combined pressure on market forces to push both the price and the rent of oil to unprecedented magnitudes. Mahdavy's ideas experienced something of a renaissance in the 1980s in the literature on the Middle East among such scholars as Hazim Beblawi and Giacomo Luciani, who were interested in the impact that this "windfall of wealth of unprecedented magnitude in such short time" had had upon the nature of the Arab states.

But if the idea of the rentier state was resuscitated, its underlying meaning was not. On the surface Mahdavy's 1970 definition appears uncomplicated and to the point. But as Beblawi and Luciani pointed out, as soon as the concept was pursued, important analytical problems emerged. To focus exclusively on the state, independently of the economy, and to define a rentier state as any one that derives a substantial part of its revenue from foreign sources and in the form of economic rent "is a rather restrictive definition that says little about the economy."[5] Beblawi and Luciani prefer instead to define the concept of a *"rentier economy"* in which rent plays a major role, and in which that rent is external to the economy.[6]

Mahdavy's definition of the *"state"* implies a meaning that is almost synonymous with that of *"society,"* indicating an entire social structure, which in the case of a rentier state is premised on the inflow of external rent. This characterization of the state by analytical conceptual abstraction (*state = "country"*) is abandoned by Beblawi and Luciani, who prefer to define the state as a political-economy: or what they call a "combination of essential indicators describing the relationship between the state and the economy."[7] This definition is chosen, they say, "not to reach an abstract notion of such a state but to help elucidate the impact of recent economic developments, in particular the oil phenomenon, on the nature of the state."[8]

For Beblawi and Luciani the *rentier state is a subset of a rentier economy*, and the nature of the state is examined primarily through its size relative to that economy and the sources and structures of its income.

Beblawi and Luciani are quick to clarify the dual meaning of the word "state"—as referring on the one hand to the "overall social system subject to government or power," while at the same time indicating "the apparatus or organization of government or power that exercises the monopoly of the legal use of violence."[9]

Nevertheless it becomes clear upon closer examination of their writings that they emphasize the latter usage over the former. For example, Beblawi uses both the terms "state" and "government" interchangeably, and Luciani refers to the state as "the structure of power and authority that exercises the attributes of sovereignty within [a country]."[10] This may reflect the reintroduction of the state as a conceptual variable in recent political literature of the 1980s.[11] It also might simply reflect the ubiquitousness of the rentier state in the domestic social order. "The state or the government being the principal rentier in the economy, plays the crucial role of the prime mover of economic activity."[12]

Beblawi delineates four characteristics which must be present in order for a state to be classified as *rentier*. First, the rentier economy—of which the state is a subset—must be one where *rent situations predominate*. Beblawi argues that there is no such thing as a pure rentier economy and concurs with Mahdavy's view that the determination of when an economy becomes rentier is a matter of judgment. Second, the *origin of this rent must be external* to the economy. In other words, the rent must come from foreign sources. Domestic rent, even if it were substantial enough to predominate, is not sufficient to characterize the rentier economy because economic rent is a factor income that only results from production (labor), investment (interest), and management of risk (profit)—i.e. internal forces of production. Third, in a rentier state *only the few are engaged in the generation of rent*, while the majority are involved in its distribution and consumption. Therefore an open economy with high levels of foreign trade is not rentier, even if it depends predominantly on rent (e.g., tourism), because the majority of the society is actively involved in the creation of wealth. Finally, the *government must be the principal recipient of the external rent* in the economy. This last characteristic is closely related to the concentration of rent in the hands of the few. It also, to use a phrase popular among contemporary political scientists, "brings the state back in" to the idea of the rentier state.

Conversely, Luciani places less emphasis on the nature or "structure" of state revenue (i.e., rent/taxes) and more on its origins or "sources" (i.e., external/internal). The key feature of a rentier state according to Luciani is that *external rent liberates the state from the need to extract income from the domestic economy.*

Mahdavy notes that the oil industry's most significant contribution is that "it enables the governments of the oil producing countries to embark on large public expenditure programs without resorting to taxation."[13] Luciani takes this state autonomy as his point of departure and proposes a new categorization which defines states by their allocative and productive functions. Unlike a "production state" that relies on taxation of the domestic economy for its income—and in which economic growth is therefore an imperative—an "allocation state" does not depend on domestic sources of revenue but rather *is* the primary source of revenue itself in the domestic economy. The policy of a production state is therefore designed to increase economic growth while an allocation state fails to formulate "anything deserving the appellation of economic policy."[14]

The primary goal of economic policy in an allocation state is spending. But beyond spending (which all states must do) Luciani breaks from Mahdavy and Beblawi and specifies conceptual boundaries for his rentier/allocation state as one "whose revenue derives predominantly (more than 40 percent) from oil or other foreign sources" and "whose expenditure is a substantial share of GDP."[14] Massive expenditure, without having to resort to domestic taxation or burdensome public debt, should have given the rentier states a short-cut to development. For Mahdavy the question is why this has not occurred:

> Perhaps one of the more crucial problems that need to be studied is to explain why the oil exporting countries, in spite of the extraordinary resources that are available to them, have not been among the fastest growing countries in the world.[15]

THEORIES OF ECONOMIC RENT

In classical economic theory, rent was understood as any surplus left over after all the costs of production had been met, and was paid to the owner of the land for use of its natural resources. Thomas Malthus described rent as "that portion of the value of the whole produce which remains to the owner of the land" and "the sole fund which is capable of supporting the taxes of the state."[16] David Ricardo understood rent as a gift of nature, which reflected both the scarce quantity and differential quality of land. In agricultural production, the rent paid to a landlord resulted from the difference between the sale price of the harvest at the marketplace and the costs of producing the harvest (i.e., wages for labor, the cost of seed, interest on capital, profit for the farmer and/or entrepreneur, etc.). If the cost of producing the harvest equaled its sale price—as might occur on marginally productive land—then in theory there would be no rent for the landlord.[17] The cultivation of inferior land might not reach a margin of rentability even when a profit was received, yet the cultivation of higher quality land usually resulted in rent. This was called "differential rent" by Ricardo and reflected the difference between the quality of marginal and intermarginal land on the macroeconomic scale.

Ricardo noted that: "Mines, as well as land, generally pay rent to their owners and this rent ... is the effect and never the cause of the high value of their produce."[18]

As with farming, the differential rent of mines reflected the physical properties of the land, but was determined by the price of the mineral in the marketplace. In neoclassical economic theory, economic rent is considered one of the four primary factor incomes of the general equilibrium (along with wages, interest, and profit). The modern microeconomic concept of economic rent is particularly useful to the theory of the mine because of its focus on the key features of the minerals industry.[19] Some mineral deposits are richer and better located than others, and therefore should produce more rent. For example, a firm might apply the theory of the mine to its evaluation of commercial viability of a particular deposit and would include, as one factor, the differential rent of this deposit versus other deposits in the company's possession. The size, location, and quality of the ore are important factors affecting the decision of when to begin extraction ... if at all: "The general rule is that the deposits for which rents are the greatest are the ones that are the most profitable to exploit now" and are determined by the price of the mineral on the present and projected future markets.[20]

In petroleum economics, the rent—or "oil surplus"— is defined as the difference between the price of a given quantity of oil sold to consumers in the form of petroleum products "and the total average cost incurred in discovering, producing, transporting, refining and marketing this crude"—which unfortunately blurs the distinction between rent, royalty, and profit in a way that classical Ricardian theory had not.[21]

Neoclassical equations that have focused on the microlevel (e.g., the level of the firm) and that have divorced rent from its social context have consequently failed to account for rent on the macro-scale. Rentier theory by no means fills this gap, but does return, for better or for worse, to the classical emphasis on macroeconomic rent and, in so doing, reintroduces the social dimension of the ownership of land and its natural resources.

By treating rent as an economic category, economic theory can tell us little about the *rentier*. For the rentier is a social agent who does not actively participate in the production process yet still shares in the fruits of the product. Leaving aside the question of the origins of rent, be it physical or social, it is clear that rent is a factor income unlike the other traditional costs of production. Wages are paid for labor, interest for capital employed, profits for the successful management of risk. For each of these factor incomes some element of sacrifice and effort is involved. But the rentier is a member of a social group that is devoid of such value added. The purest rentier is but a parasite feeding on the productive activities of others. Only nature is sacrificed.

It is interesting in this respect to recall the Lockean ideal of property and its origins in the state of nature. John Locke argues that in a state of nature the earth was common to all men who as individuals possessed a property in their own persons and therefore in their own labor. An individual acquired private property in nature through the application of his labor. "Whatsoever then he removes out of the state that nature hath provided, and left in it, he hath mixed his *labour* with, and joined to it something that is his own, and thereby makes it his *property*."[22]

Locke writes that as humankind left this state of nature (for, among other reasons, the protection of such property) the rights of property were settled by compact and agreement among communities of men. This new legal-contractual system, in combination with the invention of money, enabled individuals to exceed the "rightful bounds" of their "just property." Presumably, this was the point in history when rent collection began, along with all the other perversions of the state of nature concomitant with the social contract.

Of course such story-telling is not suggested to provide an accurate historical basis for the origin of rent. But for Locke, as for the liberal economists such as Smith and Ricardo, there are few kind words reserved for the rentier.

The rentier violates the most sacred doctrine of the liberal ethos: hard work. Is it any surprise that the classical economists were wont to factor rent as an effect of nature, or more banal, price? The rentier, absent from the value-added process, reaps a reward that does not make sense in the economic world of the Protestant work ethic.

The liberal society of Locke, Smith, and Ricardo had no place for the indolent in the calculus of wealth. Better to remove rent from the normal factor incomes and treat it as a by-product of the surplus, or worse, a vestigial remnant of the feudal order, than to accept it as part of the capitalist system itself. If the "hidden hand" of the marketplace—or even mysterious "Nature"—chose to bestow upon the rentier a gift above and beyond the normal incomes of the productive world, then who were they to question such mysterious forces? The temptation to relegate rent to agriculture and mining—two activities better associated with the *ancien régime* and thereby disassociated from the new class of burgeoning industrial wealth—must have been strong for the classical economists, concerned as they were with the logic of the coming order.

Karl Marx, however, doubts the classical interpretation of rent as the result of physical and technical differentials of the land. Moreover he is careful to distinguish between medieval rent accorded by feudal landlords, and rent received by property owners in the capitalist mode of production. Marx argues that economic rent in the former was premised on the "*use value*" of commodities, while in the latter it was the result of "*exchange value.*" Marx criticizes Ricardo for dragging the concept of economic rent "into modern time from the natural economy of the Middle Ages."[23] To Marx, rent is a *social* relation, reflective and derivative of historically specific property relations in the dominant mode of production.

A bushel of wheat tithed to a lord had as its end *use value*, whereas under capitalism that same bushel is produced for *exchange value*, in which the social relations among human beings are reified into relations among things. The nature of the rent, if any, arising from its production would have to be different. Cyrus Bina has more recently employed this Marxian notion of historical specificity in order to demonstrate that oil rent is a category of property relations unique to the capitalist mode of production:

> An element of nature, such as an oil pool, which did not have an exchange value in the previous historical epochs, has now been transformed as a necessary condition for the expansion of capital.[24]

The emergence of petroleum's "*exchange value*" is of course linked to its modern "*use value.*" But the rent derives from the latter, and is determined by who owns the oil. Bina examines the first oil shock (1973–1979) and finds that the fourfold increase in the price of oil was the result not merely of differential margins of productivity, but of the specific property relations in the oil industry which fragmented ownership of the land from the ownership of the subsurface reservoirs. The "law of capture" that operated in the United States, for example, made it possible for a legion of oil producers to acquire leaseholding rights to the surface terrain of a subsoil petroleum deposit. Under such social property relations it became increasingly rational for producers to establish petroleum-production facilities above pre-existing oil reserves where one could extract a known entity from a surface lease that may have been discovered by another producer. This resulted in the production (and over-production) of marginal fields in the lower 48 states, and set the stage for competitive rents overseas.

In the Middle East many governments were able to reduce this proprietary fragmentation, as the property relations in their societies bestowed ownership of mineral reserves to the state, which in many cases owned not only the oil reserves, but also were sovereign over much of the land above. When these governments demanded an increased price for, and greater returns from, their petroleum, it was as a result of the political leverage they possessed by virtue of their property relations *vis-à-vis* the oil reserves.

Today, most of the price of oil in the Middle East consists of economic rent. "More typically, rents comprise 95 to 97 percent of gross receipts of low-cost oil."[25] Turning Ricardo on his head, rent became the cause, and not the effect, of the high price of oil.

THE RENTIER MENTALITY

Rentier theory inherited from the classical economists the distinction between "earned" and "unearned" income. Considering the moral disdain reserved for the latter, one almost has the sense that the insidious rentier is being punished by economic laws that pass judgment on the wicked for their sins in much the same way that the Supreme Being of the Enlightenment might mete out justice on the avaricious and slothful.

Not all cultures in all times have placed the same dignity on labor. In Greco-Roman civilization the ancients extolled the man who lived on unearned income. Labor was a sordid affair fit only for slaves. Aristotle commented that hard work made man unfit to rule. In ancient Rome, the virtuous life was supposed to be one of idleness: the *sine qua non* of a liberal culture and a political career.[26]

Beblawi, however, notes that social scientists, "including economists," are suspicious of a fundamental difference between earned and unearned income. This was reflected in a deep-rooted mistrust of the economic profession against rent and rentiers. Classical economists—Malthus apart—have few kind words to say about rent and rentiers. Rentiers as a social group were assaulted by both liberal and radical economists as unproductive, almost antisocial—sharing effortlessly in the produce without, so to speak, contributing to it.[27]

The choice of the concept of the rentier economy is premised on the assumption that such an economy creates a specific mentality. The economic behavior of a rentier is distinguished from conventional economic behavior "in that it *embodies a break in the work-reward causation.*"[28] Rewards of income and wealth for the rentier do not come as the result of work, but rather are the result of chance or situation.

Mahdavy lamented this fact when he contrasted the somewhat lackluster attitude prevalent among the rentier states with the sense of alarm and urgency prevalent in most other underdeveloped countries to the massive impoverishment of the general populace, and to their conditions of economic and technological backwardness. "Whereas in most underdeveloped countries," Mahdavy writes, "this kind of relative regression will normally lead to public alarm and some kind of political explosion aimed at changing the status quo" ... in a rentier state, the welfare and prosperity imported from abroad "pre-empts some of the urgency for change and rapid growth" and may in fact coincide with "socio-political stagnation and inertia."[29]

There are two components to this inertia. The first is that the existence of relatively ample resources deludes the rentier into an expectation of ever-increasing revenues in the future. The second is that the rentier elites become satisfied with their material conditions. "Instead of attending to the task of expediting the basic socio-economic transformations, they devote the greater part of their resources to jealously guarding the status quo."[30] With high expectations that future external rents will continue to increase, spending and consumption become the order of the day. Insulated by the surrounding comforts that external rent provides, rentier elites have a proclivity to form a complacent disposition and to lack the necessity that is the mother of all invention. Beblawi argues that the break in the work-reward linkage means that, for the rentier, "reward becomes a windfall gain, an *isolated* fact."[31] Income and wealth are seen as situational or accidental, rather than as the end result of a long process of systematic and organized production.

The rentier mentality is a psychological condition with profound consequences for productivity: contracts are given as an expression of gratitude rather than as a reflection of economic rationale; civil servants see their principal duty as being available in their offices during working hours; businessmen abandon industry and enter into real-estate speculation or other special situations associated with a booming oil sector; the best and brightest abandon business and seek out lucrative government employment; manual labor and other work considered demeaning by the rentier is farmed out to foreign workers, whose remittances flood out of the rentier economy; and so on. In extreme cases income is derived simply from citizenship.[32]

Beblawi concludes that such consequences of the oil phenomenon are a "serious blow to the ethics of work."[33]

THE RENTIER ECONOMY

The most frequently cited problem with the oil-dependent economies is that, like other monoproducers, *they are highly vulnerable to external price shocks*. All oil-rentier economies have been wounded at this "Achilles heel" even if we accept that exposure to the fluctuations of price has been a shared—but not a uniform—experience, and that relative economic diversification varies considerably from one economy to the next. All states in which the oil-rent situation predominates have shared an education in the uncertainty of the world oil markets, and have taken measures to protect themselves against future trauma. Still, one has a sense that the learning curve has been short and steep for the oil-rentier states, and may not provide the whole answer.

The oil boom of the 1970s undoubtedly resulted in a spending boom in the oil-rentier states. But when the oil-price crash of the 1980s slashed their primary source and structure of public revenues (external rent), these same oil-rentiers were forced to implement austerity programs to limit the bloodletting. Even the bitter pill of austerity programs could do little to cure the rentiers of their structural and conjunctural constraints. Limiting government expenditure could not and did not alter the primary structure and source of government revenues in the oil-rentier.

Diversification is the answer for most rentier states, but what is the question? If you begin with the question, "How should a rentier state allocate its resources?" then you already deny one of the central hypotheses of rentier theory—i.e., "that the development policies and the structure of expenditure are a function of the structure of revenue."[34] There are structural constraints on what is conjuncturally possible.

If a rentier economy diversifies its sources and structures of income in a significant way, then it is no longer what it once was. How does an oil-rentier economy diversify? And diversify to what? This is where the learning curve of the 1980s flattens out. Assuming that the oil-rentier economies can purchase their development, and in so doing, become "more like us," assumes that development is a commodity rather than a process. Let us examine the mechanism of the rentier economy in order to discern the problems with development and diversification.

One problem observed by Mahdavy was that "however one looks at them the oil revenues received by the governments of the oil exporting countries have very little to do with the production processes of their domestic economies."[35] Input from local industries, including wages and salaries, payment to local contractors, and purchase of local supplies is "so insignificant that for all practical purposes one can consider the oil revenues as a free gift of nature."[36] Domestic consumption is also severely limited by the export promotion of crude. That is, since most of the oil is produced for export, little of it is left behind for local consumers.

As a result, the *petroleum industries in the oil-rentier states tend to be enclave industries that generate few backward or forward linkages.* (Backward linkages are the purchase of local inputs. Forward linkages are the domestic use of the sectoral output in further productive operations.[37]) Rentier governments have made various attempts to encourage closer integration of foreign operations with businesses in the domestic economy. They often require progressive increases in the local value-added content either within the foreign affiliate or through subcontracting to local firms. Indigenization of personnel is another avenue taken to increase local participation. But the general paucity of inter-industry linkages between the oil sector and the local economy prevent oil from becoming a leading sector in the usual way associated with certain industries in the Western industrialized economies. And as Abdel-Fadil explains: *"[T]he state becomes the main intermediary between the oil sector and the rest of the economy."*[38]

The mechanism of a rentier economy is premised on the inflow of massive amounts of external rent. This rent comes in the concrete form of foreign exchange (e.g., dollars, pounds, francs, marks, lire, yen). Access to foreign exchange is important for all developing countries because it allows them to purchase not only consumable goods (food, fuel, medicine, etc.), but also the technology and capital of the advanced industrial mode of production. Many developing countries must suffer costly balance-of-payment crises and problems with inflation to acquire these goods and services. This should not be the case with the rentier state, which exists in an economy literally saturated in foreign currency. But what seems like a blessing is in reality a curse. *The inflow of external rent on unprecedented scales throws the input-output matrix of the economy into imbalance*, as both the state and the society become increasingly dependent on the continual input of this foreign revenue.

One consequence of the large amount of external rent available in a rentier economy is that its state tends to relax constraints on foreign exchange. The availability of foreign currency in such relative abundance means that the rentier states can acquire foreign goods without the usurious costs of exchange. It also means that *imported goods have the tendency to replace domestically produced goods*, particularly in agriculture and manufacturing, which often cannot compete with foreign goods produced under economies of scale. The state purchases foodstuffs, which then compete with domestically produced foodstuffs on the local markets. Combined with the attraction of rural workers to the urban areas (where the rents are concentrated) this input-output imbalance results in a decline in agricultural production. Net exporters suddenly find themselves net importers of food, with dire consequences for those poorer groups left outside the booming oil economy.

Another consequence of the availability of large amounts of external rent is that *government can embark on big capital-intensive development projects*. Possessing the foreign exchange required to purchase foreign technology, the government has the capacity to embark on large-scale infrastructural campaigns and state-run industrial complexes. The *short-term benefits* of such programs are immense because infrastructural "development" can employ domestic labor and also because "modern" industrial complexes endow the state with prestige. The *long-term consequences* of such "modernization and development," however, are less impressive. Rather than enlarging the goods-producing capacity of the economy, inter-sectoral linkages tend to be negligible because of the high import intensity of infrastructural construction activities. The state-owned industries are often worse in that they cannot employ a significant percentage of the population and often demonstrate little commercial viability. They may even drive out small-scale local capital from similar productive activities. These state-owned industries also tend to lack backward inter-sectoral linkages, relying on constant imports for their upkeep and maintenance. When combined with the rentier mentality, the most euphoric capitalist venture may turn into nothing more than another "white elephant" as inefficiency and waste in the production processes do not halt the continued inflow of investment, nor does failure to find adequate markets for producer goods dampen the drive to "produce" ever more and more oil.

So long as "unearned" income continues to flow from the petroleum sector in unprecedented magnitude, *unprofitable but prestigious development projects may continue to enjoy government allocation*. Conversely, successful projects may lose government investment when state revenues decrease—unrelated to the success of the projects themselves. An absolute increase in industrial production may therefore exist parallel with a relative decline in the manufacturing sector without affecting the primary structure and source of revenue in the rentier economy. And the consumption of imported manufactured goods can continue at conspicuous levels without domestic production being necessary at all.

The demand for domestically manufactured goods meanwhile does not keep pace with the demand for imports. Rentiers buy more and more imported goods, and fewer and fewer domestically produced goods. There are several explanations for this phenomenon.

First there is the increased presence of what could be called "*conspicuous consumption*"—the consumption of goods for the purpose of the creation of invidious comparison.[39] The status conferred by foreign imports makes them more desirable than local imitations. This is related to the second explanation, which addresses the quality of imported goods. Not only are foreign imports desirable for the prestige associated with their consumption, but they often possess *qualitative advantages* in and of themselves resulting from the advanced-process engineering of the foreign manufacturers. Factor third is the effect of exchange rates on domestically manufactured goods. In oil-rentier economies there is a temptation for government to maintain *artificially high exchange values* for their national currencies (to facilitate the purchasing power of their money) made possible by the superabundance of foreign rent. The relative price of imported goods becomes low enough to disadvantage domestic manufacturers not only in the local economy but in the external markets. Export-oriented industry looses its comparative advantage (if it has one to begin with) and import-substitution loses its economic rationale (i.e., real profitability). A fourth explanation is that domestically manufactured goods are often produced in the *absence of viable markets* to buy them. Industries are targeted for development by government policies that have extra-market considerations in mind (e.g., employment, prestige, symbolism, diversification, etc.). Prestige-oriented industrialization of this kind is pursued for its own sake, for the perceived benefits associated with the process of industry, rather than for the real needs of the marketplace. When diversification is pursued for its own sake, the "supply" of diversified goods may not be met with effective "demand"—and *domestic industries may become net consumers rather than net producers of the national income.*

Oil-rentier states have historically served as dumping grounds for surplus production of the industrialized (oil-consuming) world. Import-substitution policies have been implemented to redress the trade imbalances inherited from the colonial (imperial) era by providing equivalent goods for the domestic markets. A desire for economic autarky may excuse inefficiency under such politically autonomous state policies. In low-absorbers with small domestic markets, such inefficiency would be the natural result of economies without scale. But the kinds of export-promotion policies which have generated the remarkably rapid rates of growth in the most successful Southeast Asian NICs mandate a comparative advantage in trade or a very competitive production of goods for export. *The oil-rentier's comparative advantage is an abundance of cheap oil*, for which the world markets demonstrate relentless demand.

But where are the rentier states ultimately going to dump their *manufactured* goods? Does the market really want more producers of such value-added goods? In both cases, agriculture and industry, the answer is no, and the effect of this unfortunate reality is the same: Rentier states suffer increased dependency on imports and great declines in their non-booming tradable sectors.

This condition, described by the IMF and the World Bank as the "*Dutch Disease*," is a pathology that has been observed in the oil-rentier economies affliced by price shocks. The Dutch Disease afflicts countries with booming oil sectors by distorting the patterns of growth in the agricultural and other tradable productive sectors of the economy.[39] The Dutch Disease takes its name from the situation in the 1970s when booming North Sea gas exports pumped massive oil rents into the Netherlands. These gas export revenues appreciated the Dutch guilder and, in so doing, exposed Dutch industries to more intense foreign competition and higher unemployment. The Dutch Disease generates sectoral reallocation of productive resources in response to a favorable price shock (e.g., 1973, 1979) when increased revenues attract further investment: "[T]he stagnation in the agricultural and forestry sectors are a result of transfers of resources, mainly capital and labor."[40] In a country suffering from Dutch Disease *the booming sector attracts rural workers away from agricultural production* while at the same time contributing to a relative devaluation of local foodstuffs. (The same happens in the industrial manufacturing sector.) Capital is re-allocated to the oil sector, where returns are higher than in either agriculture or manufacturing. Don Babai of Harvard University has noted that in the aftermath of the oil price shocks of 1973 and 1979, "the economically destabilizing repercussions of oil booms were even more pronounced than the sectoral models of the Dutch Disease would have led us to expect."[41]

Compared to the relative declines suffered by groups in the agricultural (**A**) and industrial-manufacturing (**I**) sectors, the service (**S**) sector in a rentier economy swells. Since government is the principal recipient of oil rent, *there is a tendency for bureaucracies with allocative functions to expand.* Financial services also increase to meet the needs of incoming foreign exchange. Oil service industries also experience distorted growth: pipeline maintenance, storage tanks, port facilities, helicopters and other transport businesses, and various *entrepôt* merchants supplying oil company employees in the enclaves with tertiary services and basic supplies (e.g., food).

As Ruth First suggests: "The usual development process is reversed. Instead of the progression from agriculture to industry to services, *oil provokes the growth of only the third sector.*"[42]

Why does the S-sector swell in a rentier state? One reason offered by Mahdavy is that the real exchange rate between tradable and non-tradable goods increases to the advantage of the S-sector and disadvantage of A- and I-sectors. He also cites the temptation of government bureaucracies to give pay raises and kickbacks to their employees on a regular basis.

But another explanation can be offered by focusing attention on the absorptive capacity of a rentier economy: Where is the surplus capital going to be invested? Looking at the long-term picture one might argue that it should be invested in the internal (i.e., Gabonese) productive forces. But in the short-run (the time frame in which real policy decisions are usually made) such *domestic productive forces are insufficient in size to absorb large amounts of investment* at the unprecedented rates in which they flood into the rentier economy during a boom. This explains the reallocation of external rents into foreign investment portfolios, like Swiss bank accounts, or overseas real estate, where a return on investment is more likely to accrue.

In postindustrial societies the service sector evolves to meet the needs of the other sectors. But can a service sector create industrial and agricultural sectors? In terms of actual output, the S-sector is "unproductive" and cannot replicate the productive structure of a modern industrial society nor even a traditional agricultural one. It is a classic case of placing the cart before the horse.

Labor in a rentier economy also suffers from traumatic distortion. Wages earned by some, for example, in a booming oil sector can artificially drive up wages earned by others in unrelated sectors through what Dudley Seers calls the "demonstration effect."[43] Labor is also affected by the rentier mentality, as "getting access to the rent circuit is a greater preoccupation than reaching productive efficiency."[44] Hard workers in a rentier economy often discover to their dismay that the oil enclaves pay the highest wages but that access is a matter of *whom* one knows rather than *what* one knows (or how well one might contribute to producing the national income out of the national wealth).

One must also factor in "remittance" workers who fill the labor gaps of an oil-rentier economy when the size and skill-level of the domestic labor force is limited. The presence of *migrant workers* and foreign-trained professionals in many rentier states increases noticeably during oil booms. "For the host countries," Thomas Stauffer remarks, "remittances by expatriate labour are direct costs" and should be debited from the gross national income.[45] Much of this migrant labor is low-skilled, filling the gap created by the rentier mentality in the work ethos. During boom cycles large numbers of migrant workers flood into the rentier economy in search of jobs deemed unworthy by its rentier class. During bust cycles these same low-skilled migrant workers pose a direct threat to domestic employment and suffer wholesale expulsion and xenophobic hostility.

Unionization of labor is difficult to accomplish under these conditions, not only because "remittance" workers can be emloyed and disemployed freely by the state, but also because the normal tools of influence available to labor (e.g., the strike) are ineffective in an economy where reward is not causally linked to work. In an oil-rentier state it must be remembered that petroleum reserves rather than people are exploited. "The small number of hands employed in the oil business" observes Luciani "makes it possible essentially to buy off the possibility of unions developing in that sector."[46] Also the sectoral imbalances in employment disinvest unions of their traditional source of power—i.e., factories. "[I]t is very difficult to start unionization from the service sector or from a petrochemical plant."[47]

Foreign-trained (e.g., American, French, British) professionals are not so easily expurgated nor controlled, although hostility to their presence may be periodically vociferous. In the petroleum sector they remain highly visible reminders of that industry's enclave features.

Mahdavy postulated that the negative effects of rentierism could be most clearly demonstrated by observing the postive effects of a total stoppage of oil exports. This is the theoretical *"null hypothesis." Theoretically, if massive amounts of external rent create the above distortions, then the effect of lower prices, or complete cessation of external rents, should remove such distortions.* Reducing "the supply of foreign exchange and hence of imports" actually provides a stimulation for domestic forces of production: "[B]ecause of the rise in the value of imported commodities, substitution of domestic products for marginal imports becomes profitable."[48] This suggests that the solution to the economic problems of a rentier state is *not more* government expenditure, but *less*. It also suggests that *meaningful diversification will not come from above, but from below*.

The idea of development as understood by Mahdavy assumes that industrialization is the proper measure of progress. But he wants to distinguish between apparent "prosperity" of the rentier states and what he conceptualizes as real economic "development." To comprehend his critique, it is essential to recognize his bias for the Western model of development:

> Industrialization need not of course be the only road to rapid growth. But apart from the fact that for most underdeveloped countries industrialization seems the main hope, increasing the overall productive capacity of an economy is greatly dependent on such factors as higher capital per worker, improvements in technical skills of the labour force, greater specialization and realization of potential external economies in production.[49]

In short, Mahdavy conceptualizes real economic development as a kind of significant advancement in the entire social structure, measured not merely by increased percapita income, but rather by a transformation of the social forces of production. So long as "prosperity" of the rentier states derives from external rent, technological and organizational improvements will remain *un*developed and real economic development illusory.

Perhaps the most highly visible distortion in the rentier economy is the measurement and distribution of income. The measurement of per-capita income is not a good indicator of the real material conditions of the populations surviving in a rentier economy because it includes oil rents as part of the gross national product when in fact the "black gold" is hoarded by the petroleum corporations and the rentier elites. *Real per-capita income in a rentier economy is considerably lower than the simple division of gross national revenue by estimates of the national population would have us assume.* Nafziger writes in this regard that "growth probably cannot solve the problem of widespread poverty unless attention is given to how income and property [are] distributed."[50] Rentier states may have high per-capita incomes on paper, but in reality, they have a lot of very, very poor people.

Furthermore unrequited mineral rents such as oil revenues "exercise a multiplicative impact upon the recipient economy which may be expressed as a '*rentier multiplier*'."[51] The rentier multiplier is a derivation of the familiar Keynesian concept whereby an injection of funds causes a larger increase in effective demand and leads to an increase in income greater than the net injector. External rent is injected by the government into the rentier economy through the purchase of local currency, which is spent domestically. "The recipients—civil servants, contractors, etc.—spend on imports whereby funds leak out of the economy."[52] Simultaneously (notes Abdel-Falil), "the rest of the money is re-spent at home and a second-round effect is generated, and so on."[53] The difference between the Keynesian-fiscal and Stauffer-rentier "multipliers" is considerable. The former is the product of deficit spending while the latter is the injection of unrequited mineral rents. The former multiplies the productive impetus of spending while the latter *intensifies the dependence on external rent*. The rentier multiplier also shows asymmetricality of impact, as these effects are more localized in certain sectors (e.g., construction, trade, and finance) and among certain economic groups.

When international organizations account for oil revenues as national income, rather than as a "drawdown of the capital value of the finite stock of a depletable resources," the impression of real per-capita national wealth is confounded.[54] Wealth is not being created, it is being consumed, and at conspicuous rates. Thomas Stauffer reminds us that "*the production of oil is the liquidation of a finite asset*" and must be interpreted more carefully in the determination of national income.[55]

RENTIER CLASS POLITICS

The structural problems intrinsic to the input-output imbalance of the rentier economy are mirrored in the *class structures* and political rules of the game in the rentier state. Politically speaking, the structure of public finance in an oil-rentier state tends to concentrate economic wealth—and in the process, political power—in the hands of the few. Mahdavy suggests that since oil rents are paid directly to the government, "the temptations for a government bureaucracy to turn into a rentier class with its own independent source of income are considerable."[56]

Creation of a rentier class is conditioned by the state's capacity to support itself financially through external rents and socially reproduce its authority over society. The major contribution of an oil industry to a rentier state is that it enables the government "to embark on large public expenditure programs without resorting to taxation."[57] Therefore allocative functions of government bureaucracy expand in excess of extractive capacities. Why does this occur? First of all, in the rentier state there is very little production and therefore *relatively little to tax*. Additionally, the *cost of tax-collecting administration* is high in both financial and organizational terms. But the most important factor of all is the fact that massive amounts of external rent flowing into state coffers permits the government to *purchase consent* of the governed in the short-term without paying the political price of unpopular taxes over the long-term. In the case of the rentier state, "it is better to give than to receive."

There are some social and political repercussions resulting from this absence of fiscal-tax policy. Keep in mind that the purposes of taxation are twofold: to generate government revenues, but also to redistribute wealth and income in society. Tax collection is a fiscal policy that can redirect and channel revenues in socially desirable directions. Also, despite its onerous nature, taxation builds institutions of government that enhance the capacities of the state and the economy. By liberating itself from the necessity of tax collection, the rentier state unwittingly diminishes its own administrative capacity.

Like the historical bourgeoisie of the capitalist industrial revolution, the rentier class emerges within a rentier economy and rentier state as a new social category that displaces traditional elites.[58] Their economic control of the distribution of external rent easily translates into political influence, which is magnified by the autonomy of their source of revenue.

Unlike the classical liberal third estate, however, the rentier class is not driven by any notion of the ethical superiority of work and the rightful representation based on merit. *The rentier mentality isolates position and reward from their causal relationship with talent and work.* Psychologically reinforced by the intangible nature of bureaucratic output, this rentier mentality thrives in the environment of government administration (where output is almost impossible to measure and where the rentier class is not constrained by the same bottom-line as the capitalist).[59] Popularity rather than performance determines the position of the political elite. This helps us to understand the reluctance on the part of the rentier class to collect taxes even when it possesses the political power to do so.

Usually treated as a subcategory of the service sector, the public bureaucracy of the rentier state swells to immense proportions. This swollen state apparatus has the potential for *vampirism*[60] but satisfies itself instead with more moderate parasitic behavior *vis-à-vis* the domestic society. To understand why the rentier class refuses to maximize its extractive capacity one must uncover the political "rules of the game" in the rentier state. A normal state develops a bloodsucking capacity because it needs to extract revenue from its domestic society, which in turn makes certain demands on the state in exchange for its legitimate right to collect taxes. "The state for its part must give credibility to the notion that it represents the common good" argues Luciani.[61] Tax collection therefore requires at least some degree of acceptance on the part of the population. This notion implies, in a normal state, the existence of a relationship between taxation and legitimacy captured in the American revolutionary dictum: "*No taxation without representation.*"

But in a rentier state, it is through the vehicle of public expenditure that the economy functions. Domestic society feeds on the allocation of public loans, government subsidies, construction projects, and state-planned industrial complexes. Like Dracula sucking his host, the rentier state reproduces or duplicates its own rentier mode-of-survival within the local economy. Individuals within the society become vampires dependent on state expenditure, and, contributing little to it, have little say in the way it is dispersed.

The rentier class does not need to create a national mythos giving credibility to the notion that it represents the common good. Because it spends and does not tax, the rentier class is liberated from the reliance on legitimation by its society and thus suffers no serious challenge to its power. Democracy is not a problem for the rentier state, because, to turn the phrase, there can be "*no representation without taxation.*"[62]

Given the blatant maldistribution of income and wealth, one might assume that some form of class politics would emerge in a rentier economy to redress the iniquity of material conditions. But class based politics are impossible because the economic conditions and sectoral imbalances of the rentier state discourage class formation in the usual sense of the term.

Consider the following options for class-based politics: a declining rural-agricultural sector; a state-sponsored industrial sector; a booming service sector. Whence the revolution?

In point of fact, most rentier states have institutionalized some kind of vent for popular discontent in the form of representative bodies and general assemblies. But lacking the economic clout of a taxpaying middle class, can any of these subaltern groups press their claims for increased participation? *Since the government distributes benefits, the opposition necessarily focuses its attention on how those benefits are distributed.* This shapes the entire political debate of dissent in the rentier state and produces a discourse in which "the solution of manoeuvring for personal advantage within the existing setup is always superior to seeking an alliance with others in similar conditions."[63]

Theoretically, there is no incentive for a rentier class to promote truly democratic reforms. Not only is it independent financially from such demands but also it bears a vested interest in the *status quo*. This explains the reluctance of the rentier class to engage in extractive fiscal policy (e.g., income tax). To do so would not only be unpopular, but would threaten the very security of their elite status within the power structure premised on the inflow of external rents.

For these reasons, direct redistribution of oil rents will *not* contribute to greater democracy in the rentier state, but, in fact, *will stultify it.*

CHAIN OF CAUSALITY IN A RENTIER STATE

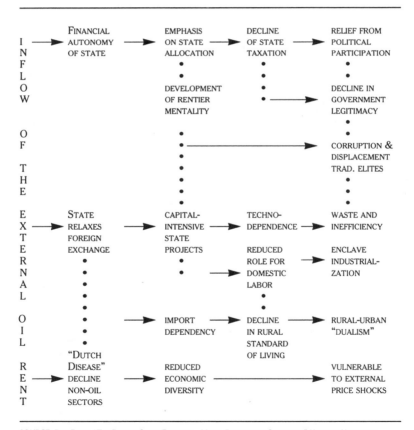

	FINANCIAL	EMPHASIS	DECLINE	RELIEF FROM
I	AUTONOMY	ON STATE	OF STATE	POLITICAL
N	OF STATE	ALLOCATION	TAXATION	PARTICIPATION
F		•	•	•
L		•	•	•
O		DEVELOPMENT	•	DECLINE IN
W		OF RENTIER	•	GOVERNMENT
		MENTALITY		LEGITIMACY
O		•		•
F		•		CORRUPTION &
		•		DISPLACEMENT
T		•		TRAD. ELITES
H		•		•
E		•		•
E	STATE	CAPITAL-	TECHNO-	WASTE AND
X	RELAXES	INTENSIVE	DEPENDENCE	INEFFICIENCY
T	FOREIGN	STATE		
E	EXCHANGE	PROJECTS	REDUCED	ENCLAVE
R	•	•	ROLE FOR	INDUSTRIAL-
N	•	•	DOMESTIC	ZATION
A	•		LABOR	
L	•		•	
	•		•	
O	•	IMPORT	DECLINE	RURAL-URBAN
I	•	DEPENDENCY	IN RURAL	"DUALISM"
L	•		STANDARD	
	"DUTCH		OF LIVING	
R	DISEASE"	REDUCED		VULNERABLE
E	DECLINE	ECONOMIC		TO EXTERNAL
N	NON-OIL	DIVERSITY		PRICE SHOCKS
T	SECTORS			

SOURCE: DON BABAI, "THE RENTIER STATE/ECONOMY—MODEL PATTERNS OF CAUSATION," HARVARD UNIVERSITY (1992).

NOTES
1. Thomas Stauffer, Hazem Beblawi, Giacomo Luciani, Mahmoud Abdel-Fadil, Michael Chatelus, Hamid Ait Amara, Dirk Vandewalle, Larbi Jaidi, Hesham Garaibeh, Afsaneh Nafmabadi, contributors to *The Rentier State*, vol. 2, *Nation, State and Integration in the Arab World*, ed. Hazem Beblawi and Giacomo Luciani (London: Croom Helm, 1987)

2. Hossein Mahdavy, "The Pattern and Problems of Economic Development in Rentier States: The Case of Iran," in *Studies in the Economic History of the Middle East*, ed. M.A. Cook (Oxford: Oxford University Press, 1970) p. 428

3. Ibid, p. 432

4. Ibid, p. 431

5. Beblawi & Luciani, *Rentier State*, p. 11

6. "External" here refers to anything originating from "outside" the spatiotemporal (space-and-time specific) boundaries of the state. These boundaries are not fixed, nor are they solid, and therefore they change according to new international realities. Synonymous with "foreign" these "external" factors are important to the discussion of any system's equilibrium (e.g. national economy, balance of trade, etc.). Specifically the kind of "external" rent we are talking about tends to come in the form of the United States dollar, which is, for historical reasons, the primary currency of international petroleum exchange world-wide.

7. Beblawi & Luciani, *Rentier State*, p. 4

8. Ibid, p. 51

9. Ibid, p. 4

10. Ibid, p. 64

11. See Peter Evans, Dietrich Rueschemeyer, and Theda Skocpol, *Bringing the State Back In* (Cambridge: Cambridge University Press, 1985)

12. Beblawi & Luciani, *Rentier State*, p. 53

13. Mahdavy, "Renter State in Iran", p. 432

14. Ibid, p. 70

15. Mahdavy, pp. 432-434

16. Rev. T.R. Malthus, *An Inquiry into the Nature and Progress of Rent* (1815), p. 1

17. In reality, there probably would not be wages for the laborer, nor profits for the entrepreneur.

18. David Ricardo, *The Principles of Economy and Taxation* (London: Everyman's Library, 1821) p. 590

19. William Spangar Pierce, *Economics of the Energy Industries* (Belmont, CA: Wadsworth, 1986). Pierce uses the phrase "theory of the mine" rather broadly: "For a comprehensive introduction to the theory of the mine that is accessible to those who know differential equations and intermediate economy theory, see Allen Kneese and Orris C. Hirfindahl, *Economic Theory and Natural Resources* (Columbus, Ohio: Charles E. Merrill, 1974). The easiest place to start on the theory is Robert Solow "The Economics of Resources or the Resources of Economics," *American Economic Review* 64, no. 2 (May 1974), pp. 1-14; the classic article according to Pierce is Harold Hotelling, "The Economics of Exhaustive Resources," *Journal of Political Economy* (April 1931), pp. 137-175; Mason Gaffney, ed., *Extractive Resources and Taxation* (Madison: University of Wisconsin Press, 1967) offers both a review of the theory and interesting empirical information; some of the voluminous recent theoretical literature is discussed in Michael G. Webb and Martin J. Richetts, *The Economics of Energy* (London: Macmillan, 1980) chap. 3 (p. 99).

20. Pierce, p. 99

21. J.M. Chevalier, "Theoretical Elements for an Introduction to Petroleum Economics" in *Market, Corporate Behavior, and the State* (eds.) A.P. Jacquemin & H.W. deJong (Hague: Martins Nijhoff, 1976)

22. John Locke, "Of Property," in *Second Treatise of Government* (Indianapolis: Hacket Publishing Company, 1980)

23. Karl Marx, *Capital, Vol. 3* (New York: Vintage, 1981) pp. 751-2

24. Cyrus Bina, "Theories of Rent and Their Critique: The Development of a Theory of Oil Rent" in *Economics of the Oil Crisis: Theories of Oil Crisis, Oil Rent, and Internationalization of Capital in the Oil Industry* (New York: St. Martin's Press, 1985) p. 47

25. Thomas Stauffer, "Income Measurement in Arab States," in Beblawi & Luciani, *Rentier State*, p. 26

26. Paul Veyne, *From Pagan Rome to Byzantium*, vol. 1, *A History of Private Life*, ed. Phillip Ariel (Cambridge: Harvard University Press, 1987) pp. 117-138

27. Beblawi & Luciani, *Rentier State*, p. 50

28. Ibid, p. 52
29. Mahdavy, p. 437
30. Ibid, p. 443
31. Beblawi & Luciani, p. 52
32. For example, in Kuwait—where Sheikh Abdullah al-Salim decided that the state should share a part of the oil rent with the population, and thereby became a distributor of favors and benefits, rather than an extractor of taxes—Kuwaitis are paid an income simply for being citizens.
33. Beblawi & Luciani, p. 62
34. Ibid, p. 8
35. Mahdavy, p. 429
36. Ibid
37. Isiah Frank, *Foreign Enterprise in Developing Countries: A Supplementary Paper of the Committee for Economic Development* (Baltimore: Johns Hopkins University Press, 1980) p. 89
38. Beblawi & Luciani, p. 83
39. Alan Gelb and Associates "Oil Windfalls: Blessing or Curse?" *World Bank Research Publication* (Oxford: Oxford University Press, October 1988)
40. Guy Marcel Eboumy, "Gabon: From Windfall Surpluses to the Crisis" (master's thesis, American University, Washington, D.C., 1989) p.14
41. From a letter to Dr. Irene Gendzier of Boston University, December, 1992, regarding a conference held on the rentier state at Harvard University that month.
42. Ruth First, "Libya, class and state in an oil economy," in *Oil and Class Struggle* (eds.) Peter Nore and Terisa Turner (London: Zed Press, 1980) p. 120
43. Dudley Seers, "The Mechanism of an Open Petroleum Economy," *Social and Economic Studies*, 13, 1964
44. Beblawi & Luciani, p.111
45. Ibid, p. 27
46. Ibid, p. 75
47. Ibid, p. 76
48. Mahdavy, p. 442
49. Ibid, pp. 435-436
50. E. Wayne Nafziger, *Inequality in Africa: Political Elites, Proletariat, Peasants and the Poor*, 2nd ed. (Cambridge: Cambridge University Press, 1989) p. 1
51. Beblawi & Luciani, p. 33
52. Ibid, p. 34
53. Ibid, p. 88

54. Ibid, p. 26

55. Ibid, p. 25

56. Mahdavy, p. 467

57. Ibid, p. 432

58. Of course there *are* cases in which traditional elites themselves become a rentier class. But even in these cases the traditional elites are transformed by the new source of income into something which they were not before. The effect of external rent profoundly redefines their relationship with the society and to one another: "The ruling patrimonial monarchies are neither traditional nor feudal" notes Luciani. "Appearances may not change much (in fact, they have changed quite a lot) but the substance is entirely different: those 'traditional' rulers today head complex and sophisticated allocation states." (p. 77)

59. As Dr. Edouard Bustin of Boston University's African Studies Center has commented, this is not entirely limited to rentier states.

60. Jonathan H. Frimpong-Ansah, *The Vampire State in Africa: The Political Economy of Decline in Ghana* (Trenton: Africa World Press, 1992)

61. Beblawi & Luciani, p. 73

62. Ibid, p. 75

63. Ibid, p. 74

Chapter 2

The Rentier
Economy in Gabon

INTRODUCTION

A rentier economy is one in which the rent situation predominates, an economy dependent on external rent. The
economy of Gabon has been dependent in two different
senses of the term. In the first sense, it has been a dependency of France, a neocolonial enclave of enduring French
interests. This has been a kind of post-imperial imperialism, or what has been called a "dominance-dependence
relationship." The second kind of dependence has been on
its primary production industries. This kind of dependence
refers to the massive influence wielded by export revenues
on the national income and public finance of Gabon, particularly from the petroleum sector. In this latter sense the
Gabonese economy has often been called "oil-dependent."
This chapter will describe the variety of means that France
has employed to control the Gabonese economy. It will also
summarize the history of the Gabonese oil industry in
order to show how oil-rent came to dominate the national
economy.

A COUNTRY DEPENDENT ON FRANCE

From the nineteenth century until the second half of the
twentieth century, Gabon existed in an imperial system

under which it was subject to French domination. One could say that under this imperial system the relations of power and wealth were elementary because real decision-making and capital accumulation in colonial Gabon were monopolized by the French. After 1910 Gabon functioned as an integrated part of the French Equatorial African Federation (AEF), which itself was an integrated part of the larger French Empire. Decolonization served to complicate these relations of power without entirely dissolving them.

After granting political independence to Gabon in 1960, as part of the comprehensive decolonization of francophonic sub-Saharan Africa carried out by the French Fifth Republic under Charles de Gaulle, the problem faced by French policymakers was how to preserve distinctly metropolitan interests (raw materials, markets, bases, etc.) without the formal machinery of an imperial structure. How would France continue to wield power over Gabon, for example, within the legal context of an international system in which both countries were now ostensibly equal?

Trade was to prove a useful tool for the perpetuation of both French dominance and Gabonese dependence. The French had deliberately and consistenly created in Gabon a trade-oriented economy with a French-oriented trade. From the outset French imperial economic philosophy was based on the concept of mutually interdependent, trade-oriented imperial economic systems. The idea was simple. The colonies were to provide metropolitan France with raw materials, in exchange for which France would provide the colonies with manufactured goods. In practice this attempt at "*inter*dependence" created an enclave export economy in Gabon that was left largely open to the outside world. This is evidenced by the fact that in 1966, six years after the end of colonial rule, Gabon still exported virtually all of the products generated by its extractive industries[1] and relied on imports for almost all of its manufactured goods.[2] The proportion of imports and exports with respect to the gross domestic product is another measure of this commercial "openness" (actually commercial dependence). In 1960 imports were 37 percent of GDP, and exports were 25 percent, meaning that trade accounted for 62 percent of total revenue in Gabon.[3] In 1972 exports accounted for 63 percent of GDP and imports 32 percent, accounting in total for 95 percent of GDP.[4] Even in 1989, more than a quarter century after decolonization, exports still were responsible for 46 percent of GDP, and imports for 22 percent.[5]

The ability of France to dominate Gabonese trade during the immediate postindependence period was facilitated by the way in which political independence was granted. Cooperation accords granted exemptions from customs duties on Gabonese exports to France and on French imports to Gabon. The accords were made a *sine qua non* of independence. In a letter dated July 15th, 1960, French prime minister Michel Debré made this point quite clear to Léon M'Ba: "Independence is granted on condition that the State, once independent, undertakes to abide by the cooperation agreements ... One does not go without the other."[6]

Looking at the direction of registered foreign trade between 1960 and 1966, for example, we find that in 1960 France accounted for 6,051,000,000 CFA francs out of a total of 11,826,000,000 CFA francs of exports; or, in other words, France purchased 51% of all Gabonese exports, almost twice the amount exported to all other EEC countries, and around twenty times the amount exported to the United States.[7] In that same year France imported 4,619,000,000 CFA francs worth of goods to Gabon, amounting to 59% of all imports.[8] By 1966 the United States had increased its share of exports to 23%, reducing France's share[9] to 43%— due primarily to U.S. Steel's involvment in the manganese industry—and did not affect the growth of French purchases in absolute terms, which amounted to 11,277,000,000 CFA francs in 1966.[10] The French also continued to dominate Gabonese imports in 1966 at 9,633,000,000 CFA francs, or 59% of the total.[11] France remains the principal trading partner of Gabon, accounting for: 45% of exports and 59.2% of imports in 1986; 41.5% of exports and 51.7% of imports in 1987; 49.3% of exports and 58.3% of imports in 1988; 46.5% of exports and 61.8% of imports in 1989; and 37.9% of exports and 60.6% of imports in 1990.[12]

Yet, as Joan Edelman Spero observes, if France continues to dominate Gabonese trade, this "trade cannot provide direct participation in Gabonese decision-making," because Gabon's principal products (wood, oil, manganese, uranium, etc.) do not need privileged markets.[13] Nevertheless Gabonese dependence on France as its main source of trade has been an important aspect of French domination of the Gabonese economy. French businesses established during the colonial period have a familiarity with the marketplace, which gives them an advantage in future commercial dealings. Also the historically evolved tastes for French goods that have been cultivated by French dominance of Gabonese imports,especially food imports, will stimulate future demand for French goods. But more important still is that if trade cannot provide direct participation, it may "work together with other more important economic means such as investment, finance, and aid to indirectly influence Gabon."[14]

French control of the Gabonese monetary system has been another method of domination of the Gabonesee economy. The establishment of the French Franc Zone in its modern form served to enmesh Gabon even more restrictively within the economic sphere of France. The Franc Zone, as it is more commonly referred to, is a monetary transaction association that includes all of France's former sub-Saharan African colonies except Guinea. It derives its name from a common currency—the CFA franc—originally created in 1945 as the franc of the "*Colonies Françaises d'Afrique*" but later changed to the franc of the "*Communauté Financière Africaine*." Gabon has been a member of the Franc Zone since independence in 1960, and along with Cameroun, Central African Republic, Congo-Brazzaville, Chad, and Equatorial Guinea (a former Spanish colony) conducts its business in the Zone through the Banque des Etats de l'Afrique Centrale (BEAC).

The association works on some basic principles. First is the free convertibility of the currency into French francs on the basis of a fixed parity of 1 CFA franc = 0.02 French franc.[15] This convertibility is unlimited and is guarantied by the French treasury. Second is the central holding of monetary reserves. All Gabonese foreign exchange is held in French francs in France in the operations account of the

French treasury. Both convertibility of CFA francs to French francs and access to foreign exchange is conducted through the operations account (*compte d'opérations*) linking the BEAC to the French treasury. "In effect," writes former U.S. ambassador to Gabon Francis Terry McNamara, "the various currencies in the franc zone constitute a single freely exchangeable money at fixed parity enjoying the backing of a common reserve held by the French treasury."[16] Through the French treasury's central holding of Gabonese monetary reserves, however, France was able to discourage and limit purchases from non-franc area countries. Gabon was required to prepare annual programs of imports from the non-franc area countries which then were formally submitted to France for approval. Later the formal submission was dropped in favor of informal discussions held in Paris. "The arrangement is made to sound informal," writes Charles F. Darlington, the first U.S. Ambassador to Gabon "but it is operated in a highly restrictive manner."[17] Eight of the sixteen members of the board of directors of the central bank (headquartered in Paris until 1972) were Frenchmen, as was the chairman of the board, who presided over the Gabonese Monetary Committee and without whose signature no decision could be implemented. Ambassador Darlington complains in his mémoires that maximum amounts were set on automobiles, radios, refrigerators, and air conditioners coming from outside the Franc Zone, "while minimum amounts are set for the same imports from France."[18] He observes that exports to the United States during this time were twice as large as imports from the United States, and suggests that France was providing Gabon with fewer dollars "than they would have if they managed their own currency affairs."[19] The suggestion is that not only was there no Gabonese national money, but that there was no Gabonese national monetary policy. This sentiment would be echoed a decade later by political scientist Joan Edelman Spero: "For all intents and purposes, there is no such thing as an independent Gabonese monetary policy; there is only a French monetary policy for Gabon."[20]

There have been some reforms in the BEAC power structure, such as the 1972 statutory revisions that moved its headquarters from Paris to Yaoundé, in Cameroun. The 1972 reforms also reduced the board of directors from 16 members to 12, and rearranged the distribution of seats so that France had only three members on the board, while Gabon increased its number of seats from one to two. Nevertheless, at its most basic, this monetary system takes extremely important decisions concerning the Gabonese economy out of the hands of Gabonese decision-makers. The most extreme example of this loss of control is the fact that neither Gabon nor any other African member of the Franc Zone was consulted before any of the five devaluations of the French franc that have taken place since 1981.[21]

Foreign aid has been another mechanism by which France conditioned Gabonese economic dependence. This foreign aid comes in many forms (aid-in-kind, financial aid, commercial aid, technical assistance) and has been a major factor in Gabonese development. Aid-in-kind often takes the form of food shipments, but in addition to such "food aid" it also includes commitment to the construction of hospitals, factories, dams, railways, or other infrastructure, which is turned over to the recipient country. Aid-in-kind can also provide a variety of services, technical materiel, and, even more common, military assistance (weapons, ammunition, supplies, etc.). Aid-in-kind can create dependence in the recipient country by providing essential goods without providing the know-how that is needed to produce such goods. A Gabonese army stocked with French weapons, for example, is dependent on future munitions sales from France. French development assistance in the construction of the Transgabonais railroad involved the usage of French gages, ensuring the continued need for French railway equipment. Not only do such "gifts" require future inputs of French materials, but when they are technologically sophisticated they also require active participation of French technical advisers. In either case, aid-in-kind promotes Gabonese dependence on France.

Financial aid, which comes in the form of grants, loans, direct investments, and program/project aid, can also foster dependence. Two key agencies have provided Gabon with almost all of its foreign financial assistance:

Fonds d'Aide et de Coopération (FAC) and Caisse Centrale de Coopération Economique (CCCE). FAC is the direct descendant of Fonds d'Investissement et de Développ-ement Economique et Social (FIDES) which was estab-lished after the war in 1946 to finance essential development projects in the colonies. The determination of what were "essential" projects was based on French colonial economic priorities and tended to focus on prof-itable improvements of the ports of Gabon. FAC tended to reinforce a bias in the Gabonese economy toward the development of export-oriented production by following in the footsteps of its colonial predecessor, FIDES. For example, between the years 1960 and 1970 the Ministry of Cooperation designated 33.4% of all FAC aid for infras-tructural development, which represented more money than was granted to rural production (14.7%), health (4.1%), education (10.8%), and general studies (2.2%) com-bined.[22] Of course, Gabon desperately needed infrastruc-tural development, but the focus of FAC funds was, for example, on the improvements of the ports of Libreville and Port Gentil, and on the skytram from Moanda's man-ganese deposits to the Congo, while no road existed between any of these cities themselves.

Deception abounds in the way the French have recorded their aid. For example, FAC aid to "rural produc-tion" in Gabon for 1960-70 totalled 17,708,000 French francs in expenditure; but out of that total, 6,000,000 were dele-gated to forestry.[23] Official FAC figures suggest that a sig-nificant proportion of aid went for "industrial production" but on closer examination this category included over nine million French francs for minerals research, and except for a study on the potential for a cellulose factory at Owendo, no direct assistance for manufacturing.[24]

France has tended to focus its financial assistance packages in grants—"88% of its total aid in fact"—which, according to economist Pierre-Claver Maganga Moussavou, is a policy that ensures France's hold on her former colonies.[25] He explains in his critical work *Does Aid Help?* that grants have ensured this hold because of the way in which they have been provided. Financial aid provided by FAC has been what is called "tied aid," which means that any Gabonese economic project receiving it—however small the French contribution—must order all of its supplies and services from French companies. Through the disguise of financial assistance, Moussavou reveals, "France finances the procurement of goods from French industries."[26]

Loans have also fostered Gabonese dependence. It may be of interest to note that before the first oil shock of the 1970s the French monetary policy toward Gabon had been to limit Gabonese public borrowing to ten percent of the previous year's fiscal revenue. After the 1973-74 price shock, however, lending policies became more lenient. Why this loosening of credit? One explanation offered by French journalist Pierre Péan is that French banks (in particular Crédit Lyonnais, B.N.P., and Société Générale) were eager to "recycle petrodollars" that were flooding into their vaults at unprecedented rates. Péan suggests that these banks aided their western clients, with the help of the French state, to sell their products to Third World countries such as Gabon, and assisted the Third World countries, through the lending of recycled petrodollars, to purchase said goods and/or services... *"épongeant ainsi le surplus de liquidités du système."*[27]

By the end of 1980 these French banks had raised 80 billion French francs of credits for exports to the Third World guaranteed by the French state, writes Péan, of which 50% was for sub-Saharan Africa.[28] It was not to those states most in need that this money was lent, argues Péan, but to those countries most capable of repaying. Nor was it important to these banks that they finance any projects in particular, but simply that they finance, and thereby recycle the surplus liquidity. Gabon met such criteria, used its natural resources as collateral, and borrowed for projects that had been deemed nonviable by the World Bank (i.e. the Transgabonese railroad).

Massive indebtedness increased Gabonese dependency in several ways. First it reduced the decision-making autonomy of the state by consuming ever greater portions of the operating budget. This can be seen by an examination of the growth of the public debt during the years 1974-86, when interest payments due each year became an increasingly onerous burden on the state budget. In 1974 total public debt amounted to 19.8 billion CFA francs, of which 14.7 billion CFA francs was principal due, and the remaining 5.1 billion CFA francs was interest. In 1980 the total public debt had risen to 114.23 billion CFA francs, of which 77.11 billion was principal and 37.12 billion interest. By 1986 public debt had increased to 147 billion CFA francs, of which 105 billion was principal, and 42 billion interest.[29] When calculated as percentages of budgetary expenses, payments for servicing the national debt represented 19% in 1974, and 35% in 1980.[30] In 1991, service of the public debt after rescheduling by the Paris Club accounted for 30% of budgetary receipts.[31]

Gabon's indebtedness reduced its autonomous decision-making even further when the precipitate drop in world oil prices in 1986 ended a period of relative financial abundance and propelled Gabon into a protracted economic crisis. The government adopted an IMF-designed austerity program which, for better or worse, reflected the loss of control it wielded over its own economy. Finally, the Paris and London Club reschedulings themselves indicate the manner in which indebtedness increased Gabonese dependency, as the state was made to swallow the "bitter pill" of austerity in order to secure future creditworthiness with Western bankers, while "[d]eep cuts in social services aggravated the poor quality of education and health services, particularly in rural areas."[32] It indicates a conflict of priorities between international financiers who demand repayment and national citizens who require services from the state: i.e., between members and non-members of the Paris Club. One might assume that an independent state would choose the needs of its citizens over the needs of foreign capital. Conversely a state that chooses the latter over the former acts like a dependency. Willingness to sacrifice the well-being of society for a good credit-rating also reflects the real interests of the Gabonese elites, who rule the country like an economic estate to be managed rather than a genuine political arena.

French technical assistance has also fostered Gabonese economic dependence. Sekou Touré's Guinea served as an example of what could happen to an ex-

colony of France that chose—in Touré's own words—
"poverty in freedom" over "riches in slavery." Guinea is
often cited as the extreme case of technical non-assistance
from a vindictive Charles de Gaulle: where all French representatives in the capital city of Conakry pulled out immediately after independence, "taking everything with them
from colonial archives and development plans to light
bulbs and the dishes in the governor's palace."[33] The case
of Gabon was quite the opposite. Léon M'Ba headed a state
in which former *colons* administered most of the important ministries, quite openly declaring his allegiance to
France. "In that period, French advisers served as coun-
sellors to all of the Gabonese ministers."[34] Ambassador
Darlington mentions M. Pigot as an adviser on M'Ba's cabinet, Etienne Raux as adviser to the Ministry of Foreign
Affairs, Marcel Vitte as adviser to the Ministry of
Education. In this light Spero reminds us that "advisers did
more than advise; they were key participants in decision-
making."[35]

Gradually the numbers of French advisers in the category of "general administration" declined, as a new generation of young Gabonese bureaucrats rose up through the
ranks to replace them. Under the regime of Omar Bongo, top
government posts were officially restricted to Gabonese
nationals. By 1978 a mere 5.4% of all the technical assistants
provided by the French Ministry of Cooperation to the
Gabonese government were in the relatively non-technical
sector of "general administration."[36] This did not reflect a
general decline, however, in the need for French technical
expertise. An examination of the official figures reveals that
between the years 1960 and 1978 the Ministry of
Cooperation provided on average 492 technical assistance
personnel to the Gabonese government.[37] This figure
steadily rose from 345 (1960) to 449 (1965) to 495 (1970) to
529 (1975) to 709 (1978).[38] The constant need for technical
assistance personnel in the General Directorates of
Economic Affairs, Finance, Customs, Direct Taxes, Data
Processing, and Price Control is indicated by the yearly
increase in their numbers after 1965.[39] The largest upsurge
in the number of technical assistants was in the education
sector, followed by infrastructure, and health.

Technical assitance provided by the Ministry of Cooperation is only part of the picture, and is usually connected with FAC aid. In addition, the CCCE provides personnel, as do various institutions such as ASECNA (air safety), BCEOM (equipment studies), OFEROM (railroads), ORSTOM (technical research), and CTFT (forestry)—and, of course, the French military, which keeps a marine infantry regiment of 800 troops stationed near Libreville.[40]

These troops are perhaps the most conspicuous form of French technical assistance. Provided by the Franco-Gabonese defense accords, and stationed in Libreville after the fall of President Youlou in the Republic of the Congo in 1963, they are an ominous reminder of France's role as the *gendarme de l'Afrique*. The Franco-Gabonese defense accords provided for both the external and the internal defense of Gabon. In the absence of any genuine external threat, their principal purpose has been to intervene in domestic civil unrest. For example, when popular discontent with the Léon M'Ba regime in 1964 resulted in an attempted coup, the French called out these troops to intervene and to protect their loyal president. Another example occured in 1990, when the troops were called out to quell civil unrest in the town of Port-Gentil. They evacuated French personnel who worked in the oil sector and stationed guards around vulnerable French oil facilities. In both cases the troops were used to protect vested French interests *in* Gabon, not Gabon itself against some external military threat.

Technical assistance personnel are required so long as Gabon lacks sufficient information or "know-how" to build railroads, dams, telecommunications systems, and so on. So long as Gabon relies on French military-technical assistants, France will continue to have a determinative role in Gabonese economic policy. French advisers occupy key positions, from which they can use their participation to advance French interests. They will do so in part because they are French, and so identify with French national interests; in part because they are members of the French civil service with at best a secondary loyalty to the country in which they are serving. Moreover, the threat of their removal remains a potent sanction in the hands of France and indicates the danger of Gabonese dependence.

Direct foreign investment has also fostered economic dependence in Gabon. A sectoral examination of the Gabonese economy will reveal that direct investments made in the form of large firms representing French positions in Gabon during the colonial period were maintained, if not strengthened, after Gabon's official independence. In fact, according to a study conducted by the General Planning Commission of Gabon in 1972, 39 percent of the large firms that dominated Gabonese productive and commercial enterprises ("value-added industries") had been established during the colonial period before the independence of Gabon was legislated into existence.[41]

In the forestry sector the Compagnie Forestière du Gabon (CFG) had been founded in 1945, and along with a handful of French firms such as Leroy-Gabon and Lutexfo-Rougier, has remained in control of the lion's share of Gabonese sylviculture. In the mining sector one can readily identify two firms: the Compagnie Minière de l'Ogooué (COMILOG), which had monopolized manganese production in Gabon; and the Compagnie des Mines d'Uranium de Franceville (COMUF), which holds a monopoly on the uranium deposits of the country. Both of these firms had been established before independence (COMILOG in 1953 and COMUF in 1958). In the petroleum sector the French oil firm Elf-Gabon has consistently dominated oil production since the founding of its predecessor Société des Pétroles d'Afrique Equatoriale Française (SPAEF) in 1945. The investment power that these firms wielded enabled private French businessmen such as Roland Bru (forestry), Henri Sylvoz (manganese), and Pierre Guillaumat (petroleum) to become directly involved in economic policy, and thereby to significantly determine the structure of the Gabonese economy and the direction of economic development.

Collectively the domination by France of Gabon's trade, monetary policy, financial aid, technical assistance, and direct investment illustrates an important feature of this country's economy, namely, that it is an economy extremely dependent on the outside world for its well-being and survival. This is important primarily because it is evidence that the nature of the Gabonese economy has been shaped by "external" forces—that is to say, by factors outside of its own control. It was designed by "external" actors during the colonial period in order to provide raw materials and was so conditioned by that colonial experience, and by the manner of its decolonization, that it remains an enclave economy unable to provide for itself without constant input from and output to the world "out there."

Giacomo Luciani identifies a key feature of the rentier state as the source of its income. State revenues must be "external" in origin, they must come from "out there." Gabon derives its state revenues from foreign sources which liberate the state from the need to extract income domestically. The dominance-dependence relationship between France and Gabon has therefore been a preconditioning factor in the emergence of a rentier economy in Gabon. Reliance on the world outside has, in effect, conditioned the state of the economy within.

It may be of interest to note that *almost* all of the rentier states identified as such in the existing literature have been former colonies.[42] This is not by any means to suggest, however, that all former colonies be identified as rentier states. In order to qualify for this haggard crew, the source of state revenues must be external in origin and the structure of these revenues must arrive in the form of economic rent. As we shall see, Gabon has not only met these specifications, but may have even exceeded them.

A COUNTRY DEPENDENT ON OIL

Oil has been to Gabon what gold has been to South Africa, what diamonds have been to Botswana, or phosphates to the Western Sahara: an export commodity by which the economy is defined. As Guy Marcel Eboumy, the *chargé d'affaires* at the Gabonese embassy in Washington, D.C., put it: "We are a country that depends on its oil, an oil-dependent country, like other OPEC members in the Middle East." For the past twenty years petroleum revenues have been the motor force behind government development plans, the gross domestic product, and the general economic climate of the country.

What are the historical origins of the oil industry in Gabon? How did this industry grow to become the single most important sector of the Gabonese economy? Why did the state in Gabon become dependent on its oil exports?

The historical roots of the Gabonese oil industry are, not surprisingly, buried in the soil of French nationalist policy. So this is where we will begin—in France, rather than Gabon. Daniel Yergin writes that "[t]he entire experience of wartime, beginning with the armada of taxis that saved Paris in the first weeks of the war, had convinced the French ... that access to oil was now a matter of great strategic concern."[43] After the San Remo Agreement in 1920 guarantied France 25 percent of the oil from Mesopotamia, the state set about finding a way to secure its share of future production. The decision was made to create a state oil company "entirely French" in terms of control.

Before the war, Georges Clemenceau was reported to have said, "When I want some oil, I'll find it at my grocers."[44] After the war one could safely say that the attitude of the French government had changed considerably. However, during the 1920s the "majors" dominated the French oil market. In her study of the emergence of state oil companies, Merrie Gilbert Klapp observes that Standard Oil of New Jersey controlled 56 percent and Royal Dutch Shell 15 percent of all oil supplies in France, together controlling over two-thirds of the French market. "This foreign domination created strong nationalistic feeling regarding French control of oil."[45]

Small wonder that French premier Raymond Poincaré refused Henri Deterding's offer for partnership with Royal Dutch Shell in the Mesopotamian adventure, preferring instead to call upon his friend, the French industrialist Ernest Mercier to find the funding necessary to launch a national champion. In a letter to Mercier dated March 1st, 1924, Poincaré stated his general objectives for the new state oil company. He wanted to develop a petroleum production "controlled by France" (he writes in this letter) to organize the development of resources "that the state holds or will hold by diplomatic accords" (i.e., the San Remo Agreement) and (most important of all to the discussion at hand) he wanted "with the support of the government, the *mise en valeur* and the *exploitation* of petroleum riches that could be discovered in France, in the colonies, and in the protectorates.[46]

Poincaré's dream of discovering oil in the French empire did not materialize through the Compagnie Française des Pétroles, even after the French government acquired a 35-percent share in CFP, because the French industrialists had originally created CFP to participate in the Turkish Petroleum Company venture in the Middle East. And besides, they were dependent on the technological know-how and experience of their Anglo-American partners, Anglo-Persian, Standard Oil of New Jersey, and Standard Oil of New York. "What started as a cooperative nationalistic effort to develop French oil capacity," explains Merrie Gilbert Klapp, "ended in a reluctant partnership between the state, French industrialists, and the Majors."[47]

In order to achieve Poincaré's vision of "exploitation of the petroleum riches that could be discovered in France, in the colonies, and in the protectorates," it would be necessary for the French state to increase its participation in the oil industry. In 1939 the government increased its control through the establishment of the Régie Autonome de Pétrole (RAP), which was 51 percent state-owned; and in 1945 the postwar government under Charles de Gaulle created a 100-percent state-owned company, the Bureau de Recherches du Pétrole (BRP). BRP set up a subsidiary branch in each of its prospective African colonies and possessions, which commenced a veritable French oil industry in Africa.

In Morocco the French created the Société Cherifienne des Pétroles, which made some of the earliest and longest-lasting (if smallest) discoveries of oil in Africa. In Algeria the BRP-backed Société Nationale d'Etudes et de Recherches Pétrolifères en Algérie (SN.REPAL) discovered oil at Hassi R'Mel and Hassi Messaoud, thereby starting the Saharan Desert oil boom. In Tunisia the firm Societe d'Etudes et de Recherches Pétrolifères en Tunisie (SEREPT) discovered oil at Cap-Bon.

Exploration for oil in French Equatorial Africa (and

thus, Gabon) began in 1947, when the BRP subsidiary Société des Pétroles d'Afrique Equatoriale Française (SPAEF) purchased a heavy rig in the United States and shipped it across the Atlantic to Port-Gentil on Mandji Island. As its name implied, SPAEF was responsible for exploration work throughout AEF, including the coastlines of Gabon and Moyen Congo. Mandji Isle was actually a peninsular outcropping frequently called the "Gabon Peninsula" on maps of the time, barely separated from the mainland by the mouth of the Ogooué river. It was chosen as the base of operations because of its strategic location between marine supply routes and the river routes into the interior of the country. Upriver on the Ogooué, Albert Schweitzer had established his world-famous hospital at Lambaréné. For years the Ogooué had been the principal inland route to an otherwise impenetrable rainforest. The importance of this forest to the Gabonese economy of the immediate postwar era cannot be overestimated. When SPAEF was first launched to search for oil, the economy of Gabon was considered a timber enclave. Before independence the *okoumé* tree furnished on average around three-quarters of total exports from Gabon; in 1957, 87 percent.[48] As *okoumé* wood was relatively light and buoyant, loggers were able to evacuate it from the hinterland via the numerous waterways that penetrated the forested zone. A popular photograph reproduced in several of the general texts on Gabon shows the port at Owendo jam-packed with logs awaiting final delivery to the sawmills of Europe. It was here at this juncture between the inland riverways and the coastal enclaves that SPAEF began its exploration for oil in Gabon.

The first wildcat wells were drilled in 1947, at Maboro 1 well located about 75 miles northeast of Port-Gentil, and at Mediela, about the same distance to the southeast. Maboro 1 was completed in April of 1948, reaching basement at 3,600 feet without success. Mediela had reached a depth of 7,000 feet by the year's end, also without success. By late 1948 a second, smaller, portable rig was brought in to assist the heavy rig with geological reconnaissance.[49] Geophysical surveys continued on the coast bordering the Gabon Basin during 1949. Meanwhile a second Maboro well was drilled, this time to a depth of 10,600 feet, still with no results. Conditions were hard on the French exploration crews, and the topography of the rainforest did not make hauling heavy rig equipment any easier. Nevertheless, after three years of drilling without a show of oil, the portable reconnaissance rig had obtained enough valuable information to encourage further deep drilling.[50] In November of 1951, drilling about 75 miles inland on the Ogooué, SPAEF reported the completion of a successful wildcat with very small production.[51] This successful completion encouraged the French government to invest SPAEF with two more heavy rigs in 1953, which began drilling in formation around Mandji Island during 1954 in order to determine if the wildcat was an isolated find or a trend. SPAEF completed six wells in 1955 about 14 miles south of Port-Gentil, with one of them flowing at a rate of 50 barrels an hour before being shut in.[52] SPAEF continued to develop the Ozouri field and made another discovery in 1956 at Pointe-Clairette, and also determined that around 100 shallow salt dome structures ran up the spine of Mandji Island.

Development proceeded rapidly during this drilling boom. Five more wells were completed at Pointe-Clairette during 1956 and two more by early 1957—all of them producers. Eight rigs were working during early 1957, producing enough small deposits to justify the installation of pumping facilities and the construction of a nine-mile-long crude pipe line from Ozouri to Cap-Lopez. In March 1957, the first tanker load of "Mandji" crude—as the blend of all crudes produced in this region would come to be called—was loaded for shipment to Le Havre, France.[53]

For the next four years the French continued to expand their initial finds, integrating Ozouri field with Pointe-Clairette field, Pointe-Clairette with Animba, Animba with M'Bega, and when a new field was discovered at Cap-Lopez, M'Bega with Cap-Lopez. In this way the entire length of Mandji Island was spanned by a pipe line that connected all of the French discoveries into one integrated network, transporting oil from the sub-surface reservoirs up the spine of the island to the port terminal at Cap-Lopez.

Production of "Mandji" crude initially averaged 3,511

barrels per day in 1957.[54] This figure was almost tripled by 1958, at 9,817 daily average barrels. The yield was from five fields, all of which were salt dome structures. Pointe-Clairette produced 3,789 bpd, M'Béga produced 3,248 bpd, Ozouri produced 2,482 bpd, Cap-Lopez produced 190 bpd, and Alewana produced 108 bpd. In total, these five fields produced more than three-and-a-half million barrels of crude oil, most of it good quality oil with viscosity ranging between 27 and 33 API.[55]

Such figures had a certain materiality about them, and they attracted the attention of two non-French majors—Mobil and Shell—to this tiny enclave of the French empire. For ten years the French had been exploring the region alone. The entrance of these two majors marks a turning point in the history of the Gabonese oil industry. After 1957 SPAEF would no longer possess a monopoly on petroleum activity within the territory. On September 30, 1958 an agreement was announced whereby Mobil Oil Française and Mobil Exploration West Africa Inc. formed a partnership with SPAEF for the prospecting and development of oil and natural gas on a 6,175,000 acre lease along coastal and off-shore zones in Gabon and Moyen-Congo (both still administered as parts of AEF). According to the terms of this partnership arrangement, the costs of exploration were to be split by SPAEF (50%) and by the two Socony-Mobil affiliates (25% each). SPAEF was however to retain operational control.

On October 1, 1958, another joint venture agreement was reached between SPAEF and Royal Dutch Shell, the latter creating a local affiliate Compagnie Shell de Recherche et d'Exploitation au Gabon (COSREG), to search for oil on a million-acre lease in southern Gabon. In this partnership COSREG, rather than SPAEF, was the operator. The motivations for this cooperative activity included the sharing of expenses and the reduction of risk of large financial losses. The partnership arrangement also reduced competition that SPAEF might otherwise have faced given the fact that much of the region was *terra incognita*, virtually unexplored, and potentially open to any foreign corporation once decolonization placed concession rights in the hands of black nationalists.

Mobil's sojourn in Gabon did not last long. Despite early successes both onshore and off-shore, the company only operated in Gabon until 1963. In 1959 a discovery had been made on the Mobil lease at the onshore Rembo Kotto field, a field that seemed to offer proof that oil did in fact exist much farther south than the French had at first suspected. But Rembo Kotto yielded high quality oil (47 API) in low quantities (340 bpd) and the French crews could therefore at best be encouraged. Rembo Kotto was located approximately 50 miles southeast of the nearest producing oil fields, so unless a larger deposit could be located, it was not feasible to develop. A large part of the Mobil-SPAEF concession was located offshore, which required more advanced technology than the French crews had been using onshore. Mobil purchased a moveable jack-up rig designed to operate in up to 100 feet of water and, in so doing, inaugurated the offshore oil industry in Gabon. On September 29th, 1961 this Mobil rig made the first offshore discovery in the country at Anguille field, located about five miles out to sea. Operated by French crews, it could be said that SPAEF's partnership with Mobil got it started in the offshore exploration business. For the next two years SPAEF and Mobil continued to prospect their concession, but without the success of their previous exploration. Eventually, for reasons which remain unclear to this day (but which may revolve around the absence of a big Middle-Eastern sized discovery), Mobil suspended its off-shore drilling program in Gabon. The government had just passed a new mining code in 1962 which provided many incentives for companies like Mobil to stay, but to no avail. The oil industry for the time being was left to the devices of SPAEF and Shell.

Shell's journey into Gabon was more lasting. Like Mobil, the Shell affiliate COSREG had good reason to be excited by the prospect of discovering oil in Gabon. In 1959 alone, crude oil production had increased abruptly by 54.3 percent, to 5,529,203 barrels, or more than 15,000 barrels per day on average.[56]

All of this production, it was true, derived from SPAEF's oil fields north of the COSREG concession, but the 1959 discovery at Rembo Kotto on the Mobil acreage proved that petroleum deposits were not concentrated in one region alone. In 1960, COSREG received a five-year exploration permit to its tract south of Setté Cama and began wildcat drilling almost immediately both onshore and offshore. According to the the French oil journal *Pétroles Informations*, Shell also conducted ten months of seismic work, including both reflection and refraction, and some magnetometric surveys in 1961.[57] After that, the company embarked on a long, intensive, and methodical exploration campaign.

Three years later COSREG reported its first major discovery. This find was on the coast at Gamba, where COSREG towed a large mobile platform from France to erect flow stations and lay submarine pipe lines and, in so doing, launched the second drilling boom in Gabon. The year was 1964, and now Gabon had two major offloading terminals: the Cap-Lopez terminal where the French pumped their "Mandji" crude; and the Setté Cama terminal where Shell unloaded the paraffin-rich "Gamba" crude. Two years later COSREG changed its name to Shell Gabon.

Political independence resulted in these kinds of nominal changes in the oil industry without changing the underlying structures of ownership and control. There was some pandering to African nationalist sentiments by SPAEF, for example, when the company changed its name in 1961 to the politically neutral Société des Pétroles de l'Afrique Equatoriale (SPAFE). This name change eliminated any mention of French Equatorial Africa and the colonial experience. But it was a purely formal and superficial change that did not in the least way alter the juridical status of the firm. Proprietary and managerial control remained in French hands. In terms of capital distribution the independent Gabonese state possessed a paltry .575 percent of the stock of SPAFE. This .575 percent had been given in 1959 to all four AEF territories (Gabon, Oubangui-Chari, Congo, and Chad) as a symbolic gift. The remaining 97.7 percent of the capital of SPAFE was held by the French. The key French proprietors were: Entreprise de Recherches et d'Activités Pétrolieres (ERAP) with a majority shareholding of 53.05 percent; Caisse Centrale de Coopération Economique (CCCE) with 14.543 percent;

Société Générale de Distribution Pétroliere (9.727 percent); Société Nationale de Financement de Recherche du Pétrole (4.692 percent); Compagnie du Nord (3.323 percent); Compagnie Française des Pétroles (1.602); and 10.723 percent held by *"actionnaires privés divers."*[58] There was little that was truly Gabonese about the Gabonese oil industry.

As mentioned above, the independent government of Gabon did revise the colonial mining code in 1962, but "much of the legal framework for foreign investment is based on French law"[59] and provided generally favorable treatment to foreign investors in the petroleum and mining sectors. This was accomplished by creating a special category (Regime I) for companies in the petroleum, timber, and mining sectors with a number of tax cuts such as reduced import duties, export taxes, indirect taxes, and turnover taxes for goods sold in Gabon.[60] The primary purpose of the Mining Code of 1962 was to encourage mining investments rather than to discourage them.

Let us consider for a moment the state of the economy at the time of political independence, in order to compare it later on with the oil-rentier state. In 1960, Gabonese wood exports of 736,673 metric tons provided more than three-quarters of all Gabonese export revenues.[61] What might be called aggressive efforts at "Gabonization" were made principally in the forestry sector. In 1961, zone 1 (the coastal productive sector) was reserved henceforth for firms owned by Gabonese nationals, and the large French firms were sold concessions in the north (zone 2) and south (zone 3) of the rainforest interior. "Gabonization" of the forestry sector was conceivable in zone 1 because of that zone's easy access to the sea and because the techniques of sylviculture in the region had been routinized over the years. But "Gabonization" of petroleum exploration was inconceivable at the time. At any rate the supposed "Gabonization" of forestry in zone 1 was slow to materialize, often resulting in nothing more than front companies, with Gabonese nationals serving as tititular heads of basically French forestry firms. "Under the cover of Gabonization what was really the cause was forestry rent," suggests Roland Pourtier in his study of the cycles of economic development in Gabon. Pourtier argues that "when all is said and done" the Gabonization of the forestry zone 1 was about "the political capacity of the state and its high personnel to reap the rewards of an activity that has not stopped being controlled by foreign enterprises."[62]

During the first years of political independence, two other natural resources—manganese and uranium—began to be extracted from the land. An examination of the country's accounts for 1960 shows that the mining sector accounted for 21.3 percent of gross domestic product, indicating the high value of these new additions to the national income.[63] But, unlike forestry, there was little hope that the government could "Gabonize" the mining sector. Like the petroleum industry, manganese and uranium exploitation in Gabon had distinctly non-Gabonese origins.

Manganese is an important mineral for the steel industry. The existence of manganese deposits in Gabon had been reported as early as 1930, but not until 1951 did a major geological expedition undertaken by the French Bureau de Recherches Géologiques et Minières (BRGM) and the United States Steel Corporation prove their immensity. The discoveries at Moanda are grand, and by present estimates represent around one-fifth of the world's known manganese reserves.[64] In 1953 a multinational (i.e., Franco-American) corporation was established—Compagnie Minière de l'Ogooué (COMILOG)—which began studies on mine organization and removal methods. The major obstacle at Moanda was the location of the deposits. The Moanda plateau was located in the Haut Ogooué region, so distant from the colonial capital city of Libreville that until 1946 it had been administered from Brazzaville as part of Moyen Congo. No roads nor trains existed between Moanda and the Gabonese ports (where the manganese would have to go if it were to be exported to world markets). COMILOG, however, rejected the idea of constructing a railroad through the Gabonese rainforest, chosing instead to construct an aerial conveyor belt—the longest of its kind in the world—from Moanda to Mbinda in the Republic of the Congo. From there the manganese ore was transported to the Congo-Ocean railroad, and down the tracks to coastal Pointe-Noire.

Uranium was discovered around the same time at

Mouana, around 25 kilometers north of the manganese deposits at Moanda, and thus creating the perception in Gabon at that time that mineral exports (rather than wood exports) were going to become the country's economic future. According to the cooperation accords signed on August 17th, 1960 between Gabon and France, uranium was classified as a strategic product. Article 4 of these accords stipulated that Gabon *"facilite au profit des forces armées françaises le stockage des matiéres et produits stratègiques"* and therefore limit or forbid their export to destinations in other countries. Article 5 requires Gabon to reserve its sale of strategic products by priority to members of the community. Combined, these articles made Gabon, in the words of French journalist Pierre Péan, *"un réservoir français de matières premières, une véritable 'chasse gardée'."*[65] The French Commissariat a l'Energie Atomique provided the initial capital through its subsidiary Compagnie Générale de Matières Nucléaires (COGEMA) which in turn owned 18.8 percent of the stock on the consortium that would ultimately develop the uranium in 1961, Compagnie des Mines d'Uranium de Franceville (COMUF).[66]

Iron ore was known to exist in the Ivindo region up north, where it had been used by the local inhabitants to make tools. Like manganese, iron ore is an essential ingredient in steel manufacture and was, like manganese, still comparatively valuable during the strong steel market of the 1960s. In 1959 the Bureau Minier de la France d'Outre-Mer (BUMIFOM) formed an iron syndicate—Société des Mines de Fer de Mekambo (SOMIFER)—to exploit their iron deposits of Ivindo. SOMIFER involved not only BUMIFOM, which held 12% of its capital stock, but also the Bank of Paris, and most important of all, Bethleem Steel, which held 50% of the syndicate's stock.[67] But since the Ivindo region was located in the northern interior of Gabon, transportation of the iron ore was for all intents and purposes impossible. When the Belinga iron ore deposits were discovered in the northeast, the same geographical difficulties were encountered. As a consequence, during the early years of independence, neither deposit was exploited, although the promise of future iron export revenues remained in the forefront of Gabonese policymakers' minds.

The vast majority of Gabonese peasants drew their living from subsistence agriculture centered on indigenous food crops like manioc, bananas, and peanuts. According to figures provided by the French Ministry of Cooperation in the 1963 development plan for Gabon, around 86 percent of the active population were engaged in agriculture.[68] More than half of all plantations in the country were two-and-one-half acres in size or less. While the other sectors of the economy grew rapidly, it was this sector that suffered the most dramatic declines. For the majority of the Gabonese population was agriculturally self-sufficient in 1960 in terms of food production. Gabon had never been an important exporter of any particular crops (coffee and cocoa plantations produced small amounts in the Woleu N'Tem region), but the greater part of its economy had existed outside the cash economy.

Several points can be made. First, although 86 percent of the inhabitants of Gabon worked in agriculture, they produced only 30 percent of the gross national product.[69] The majority of GNP came from the exploitation of timber, manganese, uranium, and oil. Second, these exports were shipped off in their raw, unprocessed state. Gabon had little manufacturing capacity of its own. Third, the companies that produced the national income were predominantly foreign. SPAFE and Shell produced the oil; U.S. Steel[70] and the other members of COMILOG produced the manganese; the French Atomic Energy Agency through COMUF produced the uranium; major French forestry companies (Rougier, Leroy, etc.) produced the lumber. Fourth, methods had been found whereby these raw materials could be extracted and exported without the need to develop any kind of internal transportation system within Gabon itself. The manganese and uranium were shipped through the Congo; the *okoumé* logs were floated down the rivers; the crude oil was pumped and piped along the coast to the tankers off-shore. All of these points support the claim that the economy of Gabon around the time of independence was an enclave economy, with few inputs from or outputs to the local society.

The major change that happened after political independence was that wood exports, although growing in absolute terms, began to represent a smaller and smaller proportion of the gross domestic product, while mineral exports and oil exports began to represent larger and larger proportions. The country changed from a wood enclave to an oil-and-mineral enclave. According to Roland Pourtier, two times more wood had been extracted from the forests of Gabon in the twenty years following independence than from 1900 to 1960, yet "while wood represented between 80 and 90 percent of colonial exports' value, it had fallen to under 10 percent since 1974."[71] The reason for this was the growth of the oil industry. Between 1960 and 1966, the percentage of the value of exports provided by wood fell from over 70% to under 40%.[72] During that same time period daily average crude oil production increased from 16,000 barrels to 28,881 barrels.[73] Proven crude oil reserves increased from 50,000,000 barrels to 500,000,000 barrels.[74] Total onshore and offshore drilling averaged over 100,000 feet per year.[75] In 1967 alone, crude oil production increased 138.6%. Responsible for this gain was Shell's Gamba field (above) which went on line in February at a rate of 40,000 barrels per day from 26 flowing wells.[76] Even though *okoumé* exports increased from 500,000 to 1,000,000 tons between 1960 and 1970, the increased production in the oil, manganese, and uranium industries began to supplant wood's relative contribution to GDP.[77]

Not only was the economy changing from one dominated by its wood enclaves to one dominated by its oil enclaves, but the oil enclaves themselves were changing from predominantly onshore to increasingly offshore concessions. A quick glance at the main SPAFE-Elf concessions between 1958 and 1974 reveals this trend: Ozouri (1958), Pointe-Clairette (1960), Cap-Lopez (1960), M'Béga (1960), N'Tchengué (1963), Batanga (1964), were onshore in the "*zone terrestre*," while Grande-Anguille (1969), Torpille (1969), Grondin Mandaros (1972), Girelle Marine (1974), Pageau Marine (1974), Barbier Marine (1974), and the exploration permits at Baudroie (1974), Mérou Sardine (1974), Mayumba-Lucina (1974), Dorée (1974) and Brémé

(1974) were all offshore in the "*zone marine.*"[78] Furthermore the previous French monopoly on oil production had been broken by Shell's Gamba concession. In 1967, when the Gamba field went on stream, Shell was at this point generating 56 percent of the total output in the country.[79] By 1969 Gabon's crude oil production, 100,000 barrels per day at the beginning of that year, was expected by some industry analysts to double by 1971 because of the Gamba field.[80] The French regained their majority standing when three new offshore fields were discovered in 1969 and quickly brought into their pipeline network. But the presence of a second producing firm increased the competitive nature of the oil industry in Gabon, particularly as new firms were given exploration leases. In 1970 Gabon experienced a 200-percent drilling increase.[81]

No longer a purely French preserve, the oil industry in Gabon became both internationalized and nationalized. It became internationalized as new foreign firms acquired exploration permits and concessions, sometimes as partners of the French, but frequently as independence operators. In 1967, for example, the Gulf Oil Co. of Gabon was established, and acquired two blocks at Iguéla and Maymba. Gulf also signed a joint venture agrement with Shell Gabon for exploration of a 1,930 square mile lease.[82] The American-based oil company had been successfully prospecting in nearby Angola, Congo, and Zaïre, but now became the first Yankee major in Gabon. Then there was the Italian state oil company ENI, which gained its own concession in 1968 (through its exploration subsidiary AGIP Mineraria) to 1,318 square miles of Gabonese territory.[83] The German firm Deminex also started operating in Gabon through some territory that SPAFE had farmed out in 1969.[84] At the beginning of 1971 Chevron-Texaco held a concession in Gabon, as did Oceanic and Union Carbide.[85] In 1972 the Ashland Oil Company had joined this group. The following year the Esso Exploration Company had acquired a 50% interest in the Ashland concession.[86] So it went, one firm after another carving up the territory into smaller blocks, inviting other companies to share the expenses and risks as partners.

The struggle for nationalization for the Gabonese oil

industry was equally dramatic. The 1970s were a time of strong nationalist aspirations in the most important oil producing countries of North Africa and the Middle East. President Bongo had demanded, in a speech commemorating the independence of his country, that all foreign companies active in his country would have to transfer their headquarters to Gabon by January 1st, 1974. President Bongo also announced that the state would take a 10-percent share in all new foreign companies that came into the country, and additionally that there would be what he called a "Gabonization of cadres."[87]

He began renegotiations with the two major oil-producing companies in which Shell and Elf-SPAFE[88] agreed to provide a 12-percent royalty on crude oil value at the well head, and to pay increased income taxes on 38 percent, as well as an additional five- to ten-percent royalty charged on offshore production (depending on the water depth of the field).[89] These renegotiations combined with the increased oil production to result in greater contributions of the oil sector to the state budget: 45 million U.S. dollars in 1970, 64 million U.S. dollars in 1971, and 74 million U.S. dollars in 1972.[90]

"Gabonization" of the country's sole oil refinery in 1973, which had been installed at Port-Gentil in 1967 by the five UDEAC member states under the title Société Equatoriale de Raffinage, resulted in a new state oil refiner: Société Gabonaise de Raffinage (SOGARA). In an interview with the Libyan daily newspaper *Al Fajr Al Jadid*, president Bongo warned the oil firms in Gabon that they would "suffer the same fate as those operating in Libya unless they respect the sovereignty and rights of the Gabonese."[91]

The reference was of course to nationalization, a threat which president Bongo never did keep. However there was a sense at this time that the concession system was already a thing of the past, a holdover from the defunct age of colonialism and imperialism, wholly inappropriate to the new age of decolonization, self-determination, and nationalism. Outright nationalization of SPAFE may not have been a realistic option for president Bongo, but the concept of "participation"—partial ownership by negotiations—was quite feasible. Therefore while Libya nationalized British

Petroleum, took over 50 percent of ENI, expropriated Hunt oil holdings, and nationalized 51 percent of other companies operating in his domain, including Armond Hammer's Occidental Petroleum; Gabon under Omar Bongo settled for a more cautious 25-percent stake in his country's two producing firms.

The single most important action taken by the Gabonese government toward nationalization and greater control of the oil industry was the application for membership to OPEC. The story goes that on a visit to Libya in the fall of 1973, president Bongo became enamoured with the bravado of Col. Khadaffi and was eager to align himself with the radical Arab states, so he suddenly and quite unexpectedly converted to Islam on October 8th of that year. "There cannot be many African countries where the number of Moslems are fewer than in Gabon" commented the British press on this sudden religious transformation, "What is the background to the conversion?"[92]

Notwithstanding the possibility that Albert Bernard Bongo had actually experienced a genuine religious experience, it is probable that both his conversion and name change (to *"El Hadj"* Omar Bongo after a visit to Mecca) were diplomatic efforts on the part of the Gabonese president to gain membership in OPEC, which was dominated by Islamic states.

President Bongo eventually did succeed in acquiring associate member status for his country in 1974. As a new team player in the politics of price, he was expected to adjust Gabon's crude oil prices so that they would fall into line with those of other cartel member states. Accordingly, the price per barrel of "Mandji" crude was raised from $2.40 in January of 1973 to $13.03 in January of 1974. Similarly, the price of "Gamba" crude was increased from $2.40 to $13.79 per barrel.[93] This veritable explosion of oil prices, petroleum economist Steven A. Schneider explains, "was not due to any physical shortage of oil" but rather was the result of "the successful use of the oil weapon by the Arab states in connection with the Middle East war of October [1973]."[94]

The year 1973-74 consequently represents a turning point in the economic history of Gabon because, by gaining membership in the OPEC cartel, President Bongo not only acquired a new identity for his country as an oil-exporting state, but he also brought Gabon into a pricing policy that would introduce the country to its first era of true prosperity. Gabon's draft budget for 1974 illustrated this newfound wealth, as the version that the cabinet submitted to the national assembly for approval represented a 30-percent increase over the previous year's public revenues. What was most remarkable to the press at that time was not only that ordinary government revenues were estimated at 40,820,000,000 CFA francs ("a rise of 8,685 m. CFA francs"), but that this massive gain had come "with no increase in taxation."[95]

In an interview with *Le Monde*, President Bongo said that he was "very satisfied" with the new agreements, and that "the increase in the price of oil" was "going to make extra income" for his country.[96] For the time being, the news was all good. Petroleum exports were generating more revenues than ever before, with oil rentals growing from $74 million in 1972, to $169 million in 1973, to a phenomenal $654 million in 1974.[97] For the first time in the history of the country the Gabonese were considered rich. At $700 per person, Gabon registered the highest per capita gross national product in all of "black" Africa (i.e., excluding South Africa) and was the third highest on the continent (after South Africa and Libya) in 1974.[98] In his message to the nation on the occasion of the country's 14th anniversary of its independence, President Bongo could proudly boast that the economy was approaching an annual growth rate of 10 percent, and that GNP was up by 30 percent.[99] So good was the year 1974 to the economy of Gabon that the government approved a draft budget of 151 billion CFA francs for 1975, which was "almost three times as large as the 1974 budget—without introducing any increases in taxation."[100]

By 1974 we are dealing with an oil-rentier economy in Gabon. Why did the economy become so? Two factors are key to the development of this kind of economy (which, it might be added, was Africa's second largest producer of uranium and the world's third largest producer of manganese at that time). First, there was a tremendous boom in the Gabonese oil industry itself, with a 20.4-percent increase in production.[101] Second, there was the soaring escalation of the oil price brought about by the first oil shock of 1973–74. Since these two factors were not static, but rather were quite dynamic, the degree to which the economy was "rentier" changed over time. Problems first manifested themselves in production-to-reserve ratios (which are used to approximate the future longevity and/or resource exhaustion of a country's oil industry). Between the years 1970 and 1975, Gabonese crude oil production grew from 108,847 bpd (1970) to 115,431 bpd (1971) to 125,220 bpd (1972) to 150,808 bpd (1973) to 202,502 bpd (1974) to 222,580 bpd (1975).[102]

As we have seen, this increased production was a great benefit to the current accounts, in that it generated ever greater government revenues, especially after the price shock of 1973–74. However, most of this production came from salt-dome structures that were typically small in size and tended to require secondary recovery processes in order to continue producing. Elf, Shell, and other petroleum companies actively operating in the country therefore were required to conduct constant drilling campaigns.

Between the years 1970 and 1974 the total footage drilled both onshore and offshore (including exploration/prospecting wells and development/production wells) kept pace with production, from 168,800 ft. (1970) to 233,723 ft. (1971) to 329,691 ft. (1972) to 347,578 ft. (1973) to 417,005 ft. (1974).[103]

But all of this drilling failed to produce any new or large discoveries, so that by 1976, when production reached an all-time high at 224,771 bpd,[104] M. Gilbert Lugol, president of Elf-Gabon, warned that Gabonese oil production had reached a "plateau" and "will start falling unless new finds are soon made and brought into production."[105] Proven crude reserves had fallen rapidly from 1,250,000,000 barrels in 1973 to less than half that amount (579,600,000 barrels) in 1975.[106]

The problem was that with production increasing and proven crude reserves decreasing, Gabon faced resource exhaustion unless foreign companies could be encouraged to continue costly exploration and uncertain discovery of new finds. The total area under exploration permits as of December 31st, 1976 was 26,429 square miles, of which Elf Gabon possessed 18,434 square miles.[107] Of the 362,647 square miles of Gabon under production license, Elf held 273,670 square miles.[108] Therefore, what Elf *did* mattered. But Elf complained that production costs in Gabon's notoriously difficult physical conditions were particularly high, estimated to be over $4 per barrel. The company also had grievances over the income tax rate of 73 percent on posted prices, even though its profit margins were "very much higher than in the Middle East, and last year averaged around $1.75 a barrel."[109] Furthermore, Elf Gabon had agreed to president Bongo's *Provision pour Investissements diversifiés* (PID) that obliged producing companies to reinvest 10 percent of their profits into industrial enterprises in Gabon. Total footage drilled in Gabon declined from 364,191 ft. to 224,142 ft. in 1977.[110] If Elf did not continue to invest in new exploration activities, and if production-to-reserve ratios remained the same, then the future of the Gabonese oil industry remained uncertain at best, and nonexistent at worst. And since oil provided four-fifths of export earnings in 1977, the future of Gabonese prosperity, so recently arrived, also seemed uncertain.[111]

President Bongo looked at these figures and suspected French complacency, so he began to aggressively seek out new, non-French petroleum corporations to participate in his country's oil industry. He invited British Petroleum, and the German firms Wintershall, Deutsche Schachtbau, and Preussag to sign production-sharing agreements with his government in 1977. These production-sharing agreements were modeled on the Indonesian state's previous efforts to increase their participation and their profits in their national oil industry without having to perform the unpleasant and hazardous task of actual nationalization. Production-sharing agreements ceded to the state a share of product at the well-head. This share could either be sold back to the producing firm at present market prices, or domestically refined and marketed by the state itself. Beneficial to the company, in that payment could be made in the form of product rather than hard currency, these agreements were also seen as beneficial to the state, which now took an active role in its country's oil operations rather than, as before, serving as a mere tax collector or (more to the point) a mere rent collector.

Production-sharing agreements required that the state create some kind of agency to conduct its supervisory and marketing roles. Like the Compagnie Française des Pétroles, a state oil company was the commonest means of ensuring the state's share of production as well as its dowmstream operations, if it had any. In 1979 Gabon established its own national oil company—"Pétrogab"—under the authority of the Ministry of Mines, "to search for and exploit deposits of oil, its transformation, and its transportation," etc.[112]

But Pétrogab did not succeed in increasing the state's role in the Gabonese oil industry. For all intents and purposes Gabon remained a rent-collector rather than a sovereign entrepreneur. In 1979, Elf Gabon was still the leading operator in the country, producing 68.1 percent of the oil exported from Gabon, and intended to remain the number one firm.[113] Elf committed itself to spending 59.6 million dollars on exploration in that year, far surpassing the paltry 4 million dollars committed by BP-Wintershall in its production-sharing agreement with the government.[114] Elf also formed a joint-venture operation with the Japanese World Energy Development Corporation (WED) which made a major offshore discovery in January of the following year. (WED is owned 50-50 by a Japanese banking group and the Japanese National Oil Company.) Elf formed another joint venture agreement with the Spanish firm Hispanoil, and obtained two more permits for itself on the northern and southern shelf offshore in 1980. By 1981, it was aggressively hunting for new discoveries in conjunction with Société Nationale Elf-Acquitaine (SNEA), Odeco, Ocean, Murphy, Gulf, Mitsubishi, Shell, and Hispanoil.

In short, Elf outmanoeuvred Pétrogab and the Gabonese government.

True, the high exploration levels began to pay back their investors, and oil reserves increased in Gabon. But the hidden cost of this success was that Gabonese participation in the oil industry remained retarded. When the state started renegotiations in 1981–82 for a 35%–41%

share in foreign firms operating in the oil industry, it only served to discourage new firms from entering the country and further solidified Elf's hegemony over Gabonese crude. Gabon became a victim of its own dependent success.

So tight was the French hold on Gabon's oil industry that only two other firms were able to become producers in the country: The first, American Oil Company (Amoco), was a Houston-based firm (originally Standard Oil of Indiana) which sent its subsidiary the Amoco International Oil Company in 1982 to explore on two off-shore leases at Inguessa Marin and Combé Marin Sud fields, and a year later discovered oil at Oguendjo field south of Port-Gentil. The second, Tennessee Company (Tenneco), a diversified comglomerate with some oil and gas interests, sent its subsidiary the Tenneco Oil Exploration and Production Company into Gabon in 1983, and discovered oil at the Obando Marin field 20 miles offshore, working with a 50% interest in the joint venture that included LASMO and Conoco (25% each). Later in 1984 Tenneco made another discovery at Octapus Marin 1.

In August of 1985 the journal *World Oil* reported that: "A few years ago, it was believed that Gabon would be the first Opec member to deplete itself out of the organization."[115] Production had peaked in 1976 at 250,000 bpd, then fell to 155,000 bpd by 1982. "But major discoveries by Amoco and Tenneco have changed the picture."[116]

What was most surprising about each of these firms' discoveries was that they were made on acreage previously relinquished by Elf. "Elf, perhaps stung by these successes and realizing that half of its offshore acreage would expire during 1985, made a major commitment to Gabon."[117] It was going to spend more money on exploration in Gabon than in any other country, according to the journal. Similarly, in the southern half of the country, where Shell Gabon acted as the operator for its French partners on a vast onshore concession, teams of petroleum geologists and drilling crews had been methodically prospecting a permit covering what was to become the find of the century: Rabi-Kounga.

Given the difficulties associated with searching through the dense equatorial rain forest which covers

most of the country, the oil potential for onshore exploration was less known than for offshore. In fact, historically, six offshore fields (Anguille, Grondin, Torpille,
Mandaros, Barbier, and Baudroie)—all owned and operated by Elf—produced the majority of Gabonese national
crude oil. Only Shell Gabon, with its Gamba and Lucina
fields in the southern half of the country, had produced
any significant amounts of oil onshore since the early days
of Mandji Island in the 1950s. In the three decades since
Ozouri, most of the new oil companies that came into
Gabon came into Gabonese waters. Exploration and production had moved offshore.

But in August of 1985, Shell Gabon teams working 87
miles southeast of Port-Gentil (the center of the oil industry in Gabon) found oil first at Rabi, then at nearby Kounga
field. According to Diana Hubbard, who interviewed Shell
officials for the *Financial Times* of London: "It took Shell
and its partner, Elf Gabon, about two years before they
began to realize that they did, indeed have a fabled `ele-
phant' on their hands."[118] An "elephant" is oilmen's jargon
for a major oil discovery of the dimensions more usually
associated with the Middle East than with OPEC's smallest member state. Early estimates of Rabi-Kounga rose to
as high as 1.2 billion barrels of recoverable crude. Rabi-
Kounga are still the largest finds in Gabonese history.

Rabi-Kounga by themselves rejuvenated the oil industry in Gabon. Output from the fields boosted Gabon's oil
production from 158,090 barrels per day in 1988 to 200,400
barrels per day in 1989.[119] The "Rabi" blend was low in sul-
phur, high in viscosity (34 API) and had production costs
low enough to discount it $1.50 a barrel to Brent North Sea
crude in 1989, the first year of its production.[120] But most
exciting of all was the fact that Rabi-Kounga seemed to
offer new evidence that oil existed in the largely unexplored rain forest that covered some 80% of the country.
"The Rabi-Kounga discovery holds great hopes for the
Gabonese economy," oil minister Julien Mpouno-Epigat
said to *Reuters* in an interview with Brian Killen: "There is
a strong chance that we will discover another Rabi-
Kounga."[121]

Production at Rabi-Kounga is divided among three partners: Shell Gabon (42.5 percent), Elf Gabon (47.5 percent), and Amerada-Hess Production Gabon Inc. (10 percent). Originally the Gabonese government possessed a 15-percent share of Rabi-Kounga, with Shell and Elf equally holding 42.5 percent; but the government share was sold in October of 1990 for 300 million dollars.[122] By 1991, Gabonese crude oil production had increased to 295,000 barrels per day, of which Shell Gabon produced 168,000 barrels (57 percent), Elf Gabon produced 104,000 barrels (35 percent), Amoco Gabon produced 14,000 barrels (5 percent), and British Gas produced 9,000 barrels (3 percent).[123]

As stated above, production figures are only half the picture of why the economy became oil-rentier. The price of oil was also determinative. Clearly the 1973 oil shock marked a turning point, in that the dramatic price increases which followed have never suffered a total 100% reversal. That is to say, oil has not sold for $2 per barrel since then, nor will it ever again be sold at that price range. However, this is not to say that oil prices have remained permanently high. On the contrary, before even discussing the nominal prices of oil over time, one should mention that the effects of the 1973 oil shock on its standard currency of trade—the U.S. dollar—immediatly amplified the dynamism of oil prices. In fact, after president Richard Nixon withdrew the U.S. dollar from the gold standard and floated it against a basket of currencies, the real price of oil and the nominal price of oil would forever require separate calculations.

Additionally, the so-called "price of oil" is actually a huge conceptual abstraction used to simplify a reality of many distinct and separate oil prices. This is not merely the case internationally, whereby Gabonese crude oil prices must be differentiated from Lybian crude oil prices which must in turn be distinguished from Saudi Arabian oil prices. It is also a recognition of the oil-price differentials within Gabon itself. At present (1994) there are seven qualities of petroleum in Gabon which correspond to differences in their chemical compositions and physical properties. In addition to the "Mandji" crude blended by Elf, there is also the "Gamba" and "Lucina" crudes of Shell, the "Rabi" blend is a fourth, in the same way that Amoco's

"Oguendjo" and British Gas' "Obando" crudes are separate and distinctly priced. The most recent additions are Elf's "Coucal" and Amoco's "Gombé," which came on stream in 1991.

The difference is more than merely nominal. For example, the official posted price of "Mandji" crude oil has always been slightly lower than "Gamba" and "Lucina" crudes, while the official posted price of "Rabi" crude was at first lower than "Mandji," then once it was blended with "Gamba" and "Ivinga," it was higher. In 1978 the official posted price of "Mandji" crude was $14.31 per barrel. "Gamba" crude was listed at the official posted price of $14.81 per barrel, and "Lucina" at the same $14.81 per barrel. By 1980 the price of "Mandji" was $31.64, "Gamba" was $32.84, and "Lucina" was $33.98 per barrel. And in 1981 these prices were $39.55, $40.77, and $41.90, respectively.[124] These figures were from January 1st of each year listed, but were by no mean static throughout those years. In fact the official posted prices changed six times in 1979 alone.

Additionally, there was a substantial difference between the official posted price (prix affiché) and the real price realized at transfer (prix realisé/prix de cession). Going back to when posted prices were introduced in October of 1973, we find that the prix affiché of "Mandji" crude was $6.00 per barrel, while the prix realisé was only $5.00 per barrel. In 1981 the differential still existed, as the posted price was $39.55 and the prix de cession was $37.00.[125] Prices moved up and down proportionately for all crudes, regardless of their absolute price differentials. But all of the above is to be kept in mind when discussing what happened to oil prices in the 1980s.

In 1986 the price of oil crashed.

In his analysis of the 1986 oil price crash, Edward L. Morse explains how overproduction of petroleum during the 1980s by new exporters (such as North Sea producers) and other non-OPEC members, when combined with quota-cheating by several OPEC members, caused the Saudis (who had acted as swing producers in order to keep prices high) to radically alter their traditionally conservative oil policy, and in the summer and fall of 1985, to "flood a weak market, with the intention of pushing prices down rapidly."[126]

This was a gamble, according to Morse, by the Saudis, to regain their market share, punish quota-cheating, drive out uncompetitive non-OPEC producers, and most important to regain lost revenues from their oil which had declined with decreased oil production from $113 billion in 1981 to $28 billion in 1985.[127]

The consequences of the oil price crash on Gabonese rentierism are our primary concern. For if Gabon seemed to have resolved its problem of resource depletion, there was little or nothing the Gabonese could do to palliate the free-fall of oil prices after 1986.

First we need to look at what happened to the several benchmark crudes that Gabon exported. "Mandji" crude quickly declined from $31.71 on January 1st 1985, to $24.92 on January 1st 1986, to $14.72 by July 1st 1986.[128] "Gamba" crude similarly fell from $33.64 to $14.84 during that same time period[129]; "Lucina" from $32.56 to $15.23.[130] By November of that year, crude export earnings were officially forecast to fall almost 60% to just under $900 million, which raises a second consequence: oil revenues.

Fuel export revenues (when valued at current prices) fell from $1,956,000,000 in 1980 to $886,000,000 in 1987 and then even further to $740,000,000 in 1988.[131] The fall in oil revenues raises our third concern: the relationship between oil and the overall economy.

According to the *Mining Annual Review*'s Bernard Grenier: "The Gabonese economy is based essentially on oil, which accounts for 43% of GDP, 83% of export revenues, and 64% of budgetary receipts."[132] But with the fall in prices, the contribution of oil to export revenues fell to 65%.[133] By June of 1991 the *Mining Annual Review* was reporting that oil accounted for "approximately 31 percent of GDP."[134] It may be important to note, in this regard, that during 1991 the Gabonese National Accounting System was revised to conform with the United Nations Accounting System, which resulted in a change in the method of calculating GDP for 1988-1991. "This new method revalues the level of activity (rents and services) of certain companies," comments Maison Lazard in its annual report on the country.[135] According to this accounting firm Gabon experienced a nominal decrease of the relative contribution of the oil sector to GDP from 35.2% in 1990 to 33.4% in 1991.[136]

The point is that by the 1990s the economy of Gabon was relatively less rentier than it had been during the 1970s and 1980s, because of the loss of oil revenues.

A fourth consequence of the price crash of 1986 was a negotiated agreement between the government of Gabon and the companies that set an official selling price for oil, thereby eliminating the differentials between posted and actual prices. For four years the price of "Mandji" crude remained stable at $17.32 per barrel; meanwhile "Gamba" and "Ivinga" and "Rabi" eventually settled at $17.42 per barrel; "Lucina" held at $17.77 per barrel; and "Oguendjo" at $17.50 per barrel.[137] Under the new arrangement the government benefitted from relatively secure offtake by the operators, who in turn benefitted from stable prices in line with market developments.

But neither the state nor the corporations could "hold back the sun." Oil prices fell in 1991, with benchmark "Mandji" crude dropping to as low as $16.36 per barrel. Eventually Gabonese crude oil prices regained lost ground in 1992 when "Mandji" sold for $16.74 per barrel, "Lucina" at $18.76 per barrel, "Oguendjo" at $17.66 per barrel, and "Rabi" (which now signified the blend of all the major Shell crudes—i.e., Gamba, Ivinga, and Rabi-Kounga) at $18.66 per barrel.[138] Only during the Gulf War between the U.S.-led coalition forces and Iraq did oil prices rise above $20 per barrel, and then only briefly.

The end result of recovering oil production and declining oil prices proved to be a mixed bag of goods. Looking at GDP breakdown by economic sector (as provided by the Direction Générale de la Statistique et des Etudes Economiques), we see that petroleum accounted for 208.4 billion CFA francs out of a total GDP of 1,140.8 billion CFA francs in 1988.[139] This means that oil in real terms represented a mere 18.26% of GDP two years after the fall. By 1991 petroleum generated 512.2 billion CFA francs (33.4%).[140]

If these figures appear to be too small to warrant calling contemporary Gabon a rentier economy—at least an oil-rentier economy—one must also understand that exports of crude oil and some derived/refined products (from the SOGARA refinery) altogether represented 81.6% of total exports in 1991. Furthermore, petroleum rents represented 53.7% of total state revenues.[141]

This does lend some support to the notion asserted by Luciani and Beblawi that a rentier economy naturally tends to produce a certain kind of state: i.e., one that is significantly dependent on rent. For even when oil's contribution to GDP started to decline, its contribution to public finance remained robust. In 1986 oil accounted for 63% of government revenue,[142] in 1987 it accounted for 65%[143] or 70%[144] (depending on the source), and in 1988 accounted for 70%.[145] The development of an oil-rentier economy appears to have preconditioned the development of an oil-rentier state. "Without fundamental redirection," the World Bank warned, Gabon will remain "an enclave economy, driven by the immediate consumption of oil rents."[146]

NOTES

1. Gabonese Service National de la Statistique et des Etudes Economiques, *Annuaire Statistique, Années 1968 et antérieures*, chap. 9, pp. 7-9.
2. International Monetary Fund [IMF], *Cameroon, Central African Republic, Chad, Congo (Brazzaville), and Gabon*, vol. 1 of *Surveys of African Economies* (Washington, D.C.: IMF, 1968), p.329.
3. Hugues Alexandre Barro Chambrier, *L'Economie du Gabon: Analyse Politique d'Ajustement* (Paris: Economica, 1990), p. 39.
4. Ibid.
5. Ibid.
6. Pierre-Claver Maganga-Moussavou, *Economic Development—Does Aid Help? A Case Study of French Development Assistance to Gabon* (Washington, D.C.: African Communications Liason Service, 1983) p. 166.
7. IMF, *Surveys of African Economies*, 1:331.
8. Ibid., pp. 331-332.
9. Ibid., p. 330.
10. Ibid., p. 331.
11. Ibid., pp. 330-331.
12. Annual report on the Economy and Finances of the Republic of Gabon prepared by Maison Lazard et Compagnie (1992): p.32.

13. Joan Edelman Spero, *Dominance-Dependence Relationships: The Case of France and Gabon* (Ph.D. dissertation, Columbia University: 1973), p.158.

14. Ibid.

15. This year France devalued the CFA franc to .01 French franc, or, 100-to-1.

16. Terry Francis McNamara, *France in Black Africa* (Washington D.C.: National Defense University Press, 1989), p.111.

17. Charles F. and Alice B. Darlington, *African Betrayal* (New York: David McKay Co., 1968) p.112.

18. Ibid., p.113.

19. Ibid.

20. Spero, *Dominance-Dependence*, p.231.

21. For first four devaluations, see Bernard Vinay, Inspecteur Général des Affaires d'Outre-Mer, "La zone franc d'aujourd'hui," *Marchés Tropicaux* (November 28, 1986):2981.

22. Secrétariat d'Etat chargé de la Coopération, FAC 1960-1970; cf., Spero, *Dominance-Dependence*, p. 227.

23. Spero, *Dominance-Dependence*, p. 228.

24. Ibid., p. 213.

25. Maganga Moussavou, *Economic Development*, p. 58.

26. Ibid., p. 68.

27. Pierre Péan, *L'Argent Noir: Corruption et Sous-développement* (Paris: Editions Fayard, 1988), p. 105.

28. Ibid., p. 107.

29. Chambrier, p. 335.

30. Ibid., p. 145.

31. Ibid.

32. World Bank, *Trends in Developing Countries*, p. 212.

33. Sanford J. Ungar, *Africa: The People and Politics of an Emerging Continent* (New York: Simon & Schuster, 1989), p. 358.

34. Darlingtons, p. 216.

35. Spero, p. 216.

36. Maganga Moussavou, p. 125.

37. Ibid., p. 122.

38. Ibid., p. 123.

39. Ibid., pp. 125-129.

40. International Institute for Strategic Studies, *The Military Balance* (London: IISS, 1992), p. 132.

41. Maganga Moussavou, p. 149.

42. Saudi Arabia, for example, never was a colony, strictly

speaking.
43. Daniel Yergin, *The Prize: The Epic Quest for Oil, Money & Power* (New York: Simon and Schuster, 1991), p. 189.
44. Ibid., p. 189.
45. Merrie Gilbert Klapp, *The Sovereign Entrepreneur: Oil Policies in Advanced and Less Developed Capitalist Countries* (Ithica: Cornell University Press, 1987), p. 178.
46. Jean Rondot, *La Compagnie Française des Pétroles: du franc-or au petrole-franc* (New York: Arno, 1977), pp. 31-32.
47. Klapp, p. 179.
48. Marc Aicardi de Saint-Paul, *Gabon: The Development of a Nation* (London: Routledge, 1989), p. 48.
49. *World Oil* (July 15, 1949):923.
50. *World Oil* (July 15, 1950):200.
51. *World Oil* (August 15, 1953):227.
52. *World Oil* (August 15, 1956):38.
53. *World Oil* (August 15, 1957):32.
54. *World Oil* (August 15, 1958):204.
55. *World Oil* (August 15, 1959):203.
56. *World Oil* (August 15, 1960):156.
57. *World Oil* (August 15, 1961):163.
58. Roland Pourtier, *Le Gabon: Etat et Développement* (Paris: Harmattan, 1989), p. 195.
59. U.S. Department of Commerce, *Report on Economic & Commercial Climate of Gabon* (Washington, D.C.: USDOC, 1991), p. 23.
60. This special status was accorded in all the UDEAC member states, as described in the 1968 IMF study, *Survey on African Economies, op. cit.*
61. David Gardinier, *Historical Dictionary of Gabon* (London: Scarecrow Press, 1981), p. xxiv.
62. Pourtier, p. 146.
63. Maganga Moussavou, p. 9.
64. James Barnes, *Gabon: Beyond the Colonial Legacy* (Boulder: Westview Press, 1992), p. 83.
65. Pierre Péan, *Affaires Africaines* (Paris: Fayard, 1983), p. 43.
66. Barnes, p. 84.
67. Ibid.
68. Brian Weinstein, *Gabon: Nation-Building on the Ogooué* (Cambridge: MIT University Press, 1966), p. 200.
69. Ministère de la Coopération, République Française, *Economie et Plan de Développement: République*

Gabonaise (Paris, 1963), p. 8.
70. U.S. Steel controlled 49% of COMILOG in 1960 (Pourtier, p. 199).
71. Pourtier, p. 145.
72. Ibid., p. 147.
73. "World Roundup Report," *World Oil* (August 1961, 1967).
74. "World Roundup Report," *World Oil* (August 1961 and 1967).
75. "World Roundup Report," *World Oil* (August 1961 thru 1967).
76. "World Roundup Report," *World Oil* (August 1968).
77. Pourtier, p. 147.
78. Elf Gabon, *Annual Report 1977* (Paris: 7 rue Netalon) p. 7.
79. "World Roundup Report," *World Oil* (August 1968).
80. "World Roundup Report," *World Oil* (August 1969).
81. "World Roundup Report," *World Oil* (August 1971).
82. "World Roundup Report," *World Oil* (August 1968).
83. "World Roundup Report," *World Oil* (August 1969).
84. "World Roundup Report," *World Oil* (August 1970).
85. "World Roundup Report," *World Oil* (August 1971).
86. "World Roundup Report," *World Oil* (August 1973).
87. *West Africa* (September 17, 1973):1311.
88. The adoption of the "Elf" title came about with the creation of *Essences et Lubrifiants de France*, the trademark under which ERAP (which took over BRP and other state oil interests in 1965) petroleum products were marketed.
89. "World Roundup Report," *World Oil* (August 1972).
90. Gardinier, p. xxv.
91. *West Africa* (October 29, 1973):1542.
92. *West Africa* (October 8, 1973):1420-1421.
93. Chambrier, see *Annexe I: Evolution des prix pétroliers* at end of text, *op. cit.*
94. Steven Schneider, *The Oil Price Revolution* (Baltimore: Johns Hopkins, 1983), p. 105.
95. *West Africa*, November 13th, 1973, p. 1646.
96. *West Africa*, February 25th, 1974, p. 218.
97. Gardinier, p. xxv.
98. *West Africa*, "Gabon leads Africa in GNP Growth Rate" (April 8, 1974):408.
99. *West Africa* (August 26, 1974):1052.
100. *West Africa* (October 7, 1974):1225.
101. "World Roundup Report," *World Oil* (August 1975).
102. "World Roundup Report," *World Oil* (August 1971, 1972, 1973, 1974, 1975, and 1976).

103. Ibid.
104. Ibid., 1977.
105. *West Africa* (December 13, 1976):1889.
106. "World Roundup Report," *World Oil* (August 1974 and 1976).
107. "World Roundup Report," *World Oil* (August 1977):153.
108. Ibid.
109. *West Africa* (December 13, 1976):1889.
110. "World Roundup Report," *World Oil* (August 1976 and 1978).
111. Economist Intelligence Unit, *World Outlook '77*; cf., *West Africa* (March 21, 1977):561.
112. *West Africa* (August 13, 1979):1470.
113. "World Roundup Report," *World Oil* (August 1979).
114. Ibid.
115. "World Roundup Report," *World Oil* (August 1979).
116. Ibid.
117. Ibid., p. 114.
118. Diana Hubbard, "Gabon's Oil Refuses to Run Out," *Financial Times* (August 24, 1989):26.
119. "World Roundup Report," *World Oil* (August 1990):96.
120. Maison Lazard, *op. cit.*, p. 11.
121. Brian Killen, "Gabon's Jungle Oil-Field Brings Hope to Sagging Economy," *Reuters* (May 4, 1989).
122. Xinhua General News Service, "Gabon Sells Oil Field Share to U.S. Company," October 19, 1990.
123. Maison Lazard, p. 11.
124. Chambrier, *Annexe I, op. cit.*
125. Ibid.
126. Edward Morse, "After the Fall: The Politics of Oil," *Foreign Affairs* (Spring, 1986):794.
127. Ibid:795
128. Chambrier, *Annexe I.*
129. Ibid.
130. *Platt's Oilgram News* (November 24, 1986):4.
131. World Bank, *Trends in Developing Countries*, p. 216.
132. Grenier, "Lettre Afrique Energies" *Central Africa*, p. 411.
133. *Platt's Oilgram News, op. cit.*
134. *Mining Annual Review*, 1991, Central Africa, p. 119.
135. Maison Lazard, p. 6.
136. Ibid.
137. Maison Lazard, p. 14; *cf., Direction Générale de l'Exploration des Hydrocarbures.*
138. Ibid.

139. Note: This is after the revaluation according to the new
    Gabonese National Accounting System.
140. Maison Lazard, p. 6.
141. Ibid., p. 9.
142. "Leader of tiny, oil-rich Gabon puts his nation on finan-
    cial fast track," *Christian Science Monitor* (December 4,
    1986):26.
143. *Mining Annual Review*, June, 1987.
144. Lucien Minko, "Gabonese Leader Urges to Break with Oil
    Dependency," *Reuters* (August 17, 1987).
145. "A Changed Gabon Fetes 28th Anniversary of
    Independence," *Reuters* (August 15, 1988).
146. World Bank, *Trends in Developing Countries*, p. 213.

Chapter 3

The History of the Gabonese State & the Bongo Regime

INTRODUCTION

The purpose of this chapter is to discuss the historical people and events that have shaped the state in Gabon and are directly linked to its present political regime under His Excellency El Hadj Omar Bongo. Theoretically the development of an oil rentier economy could have resulted in any number of authoritarian regimes, given the state's liberation from the society over which it reigns supreme. Historically speaking, however, the Gabonese state under president Bongo is a specific kind of dictatorship that is almost caricatural in its exaggerations of dependency and personal rule. This is a political history of Gabon. It asks the question of how the process of state formation resulted in the specific political configuration of actors who presently rule the country. Then it assesses the prospects for democratization and political change.

THE MAKING OF THE STATE IN GABON

Romanticizing the history of the Gabonese state is not the

purpose of this chapter, nor is it to idealize the state itself. The regime of president Omar Bongo has been described by political scientists Robert Jackson and Carl Rosberg as an "African autocracy," or an "absolute government by right" in which "administration has largely displaced politics as the principal method of government."[1]

For those who love democracy as the preferred choice of public order, there is little to idealize about the Bongo regime. "It is only a slight exaggeration" write Jackson and Rosberg "to suggest that the `state' under autocracy of the African type is more the ruler's private domain than the public realm."[2]

Such descriptions of the state in Gabon raise serious questions as to what exactly is meant by the term *state*. If Jackson and Rosberg's approach is valid, then "[i]n important ways the country remains an economic estate to be managed rather than an arena of politics."[3] The state from this perspective is but an instrument of its autocrat, who "conducts himself and is treated like the proprietor of the state."[4] Personification of the state under the personal rule of president Bongo is a popular idea because 26 of Gabon's 33 years of independence have been dominated by the political leadership of this one man. It can be hard to differentiate the state in Gabon from the individual who rules over it.

In contrast to the "personal rule" approach of Jackson and Rosberg which paints a portrait of the Bongo regime as an absolutist state with autonomous powers of decision-making, there is another "neocolonial" approach which paints a less potent image of the Bongo regime. French journalist Pierre Péan is one of the most notorious advocates of this school of thought, going as far as to call the Bongo regime "an extreme case, verging on caricature, of neocolonialism."[5] This "neocolonial" approach to the state in Gabon basically describes president Bongo as a puppet of the French state—in particular, of certain rightist elements within the French state associated with the infamous Jacques Foccart.

We have already discussed the legion of methods employed by the French that enable them "to participate in a definitive and determining way in the decision-making process" of Gabon.[6] But according to Péan, Gabon ceased being merely an "excrescence of the French Republic" in 1974 and became what he satirically called "Foccartland"—ruled conjointly by Bongo, his right-hand man Georges Rawiri, French ambassador to Gabon Maurice Robert, Elf-Gabon chief Maurice Delauney, French mercenaries trained by security specialist Pierre Debizet, and other members of the shadowy network of ex-soldiers and spies known as the "Foccart network" (*le réseau Foccart*).[7] From this perspective it can be hard to differentiate the Gabonese state from the French state which rules over it.

What both of these popular approaches to the Bongo regime have in common, despite their ostensible differences of perspective, is a shared critique of the existence of a classical-Weberian "state" in the country. What I mean by a classical notion of the state is an incorporation of elements of a legal order, a bureaucratic administration, jurisdiction over a territory, and the "monopolization of the legitimate use of force."[8] This is the Weberian "modern state" typology, and the "personal rule" approach negates its existence in Gabon by suggesting that real political power rests in the hands of an individual rather than in institutions. The "neocolonial" approach also negates its existence by suggesting that real power rests in the hands of the former colonial metropole. Both suggest that there is no "state."

While I am not interested in providing a narrow or restrictive definition of the state, I am interested in explaining why the state in Gabon has been so easily dismissed. In order to do this it is necessary to begin in the beginning, as it were.

Long before the classical European typology of the modern state, long before the arrival of the European on the central African coast, there were "states" in Gabon. Most historians, for example, now agree that there were kingdoms present in the Gabon estuary before the fifteenth century. We can read in Barent Erikzoon's 1594 adventures that the Gabon estuary "hath three mightie Kings in it, which raigned therein," and that "he of Pongo is the strongest of men, and oftentimes makes warre upon him of Gabom [sic]."⁹ Karl David Patterson has performed extensive historical research on the Orungu monarchy which descended from this legendary Mani Pongo, and he claims that Pongo's was the "only real state on the entire coast between the Cameroon River and the northern marshes of the Loango kingdom at Mayumba and Setté Cama."¹⁰

Dense equatorial rainforests were inimical to the development of large political entities. As a consequence, with the exception of the larger coastal kingdoms such as the Orungu monarchy, most political units tended to derive from small communities based on hunting and fishing, or limited pastoral or agricultural modes of production. Nicolas Métengué N'Nah has described the basic unit of political organization in precolonial Gabon as the clan. These clans were associations ruled by authoritative decision-making hierarchies comprised of individuals bonded together in collective leadership or kinship. N'Nah describes political life in precolonial Gabon as the organization into "a fraction of a clan, a clan, or a group of local clans."¹¹

Clans essentially fell into three categories of complexity and size when it came to their political function. The first was the *village-Etat* ["village state"] which was a small independent political entity with control over its surrounding territory, ruled by a village chief who served as a spokesman for the village council of elders. The second was the *régime confédéral* ["confederation"], which submitted each village chief to the authority of a superior confederation chief and a confederation council of elders. The third was the *organisation en royaume* or "kingdom," which was ruled by a monarch proper who held real and supreme authority over his entire domain. A less restrictive definition of the state—one that accepts these examples as valid political forms—permits us to look at these early polities as more than just quaint antiquities. Today these African states are gone. But they attest to an alternative path of political order not taken. Their basis was destroyed.

Of all the forces that served to destroy these ancient kingdoms, the most insidious was slavery. Introduced by the Portuguese in the sixteenth century, the slave trade established at Cap-Lopez grew to meet the increased demand for labor in the New World, and by the eighteenth century had become the dominant trade in the Estuary. It is estimated that between the years 1780 and 1815 an average of 500 to 1,500 slaves per year were shipped off from Cap-Lopez to São Tomé and Príncipe, and from there to the Americas.[12] The corrosive effects of slavery were twofold on the precolonial African states: They served (1) to depopulate, and (2) to cooptate. The direct victims of enslavement suffered from having their productive youths captured and transported, thereby retarding their social reproduction. In addition, at least some of the precolonial kingdoms became *comprador* slave states, such as the Orungu monarchy, which by the nineteenth century had become an importer of slaves from the interior. As Patterson explains, "[o]nce such a system was established, the coastal people could leave the dirty work of procurement to others."[13] In this way the corrupting influence of the slave trade set in motion a domino effect that destabilized the entire region.

The Treaty of Vienna (1815) officially abolished the slave trade, and brought the French navy into central African waters, ostensibly to enforce the treaty's provisions. The French, also motivated by the prospect of expanding their trade, and eventually carved out a sphere of influence for themselves in Gabon and the surrounding regions via the three expeditions of Pierre Savorgnan de Brazza (1875–1878, 1879–1882, 1883–1885). His contribution to the "Scramble for Africa" was the establishment of the Colony of the French Congo in 1886, which introduced French sovereignty in the middle and upper Ogooué river system as provided for in the negotiations of the Berlin Conference (1884–1885).[14] French colonial policy destroyed traditional African states by reducing them to mere subordinate units of the French empire. The Colony of the French Congo integrated the territories of hundreds of tiny village-states, as well as the larger confederations of villages and the kingdoms of the precolonial political order. Although slavery had been abolished, the combination of racism and paternalism was responsible for the birth of a new social order ruled by and for expatriate French tyrants.

Tyranny in its present form probably owes more to the nature of foreign domination of Gabon than it does to any traditional social order. Perhaps the worst chapter in the history of foreign rule in Gabon was the "Concessionaire Regime of 1899" when the colonial administration under Henri de la Mothe (governor 1898-1901) granted practically unregulated rights to around forty chartered concessionaires over the indigenous peoples of the territory. It was during this dark passage in Gabonese history that the "red rubber" trade flourished in the French Congo. In their lust for quick profits, civilized men turned savage, and their business practices took on the characteristics of a "feral" capitalism.

Like those described by Joseph Conrad in his *Heart of Darkness*, the Gabonese concessionaires were tyrants of Kurtzian anomie. There are many accounts of the "red rubber" trade. One of these is recorded in the journals of the renowned humanitarian, Dr. Albert Schweitzer, who describes the inhuman practices of the concessionaires and their regime:

> It was before the period of rubber plantations, so the only source of supply for world trade was the rubber extracted from the lineas or vines of the jungle. But the search for it was hard and disgusting work. That natives engaged in it were obliged to spend many long weeks in the forest and swamps, to endure famine, because being far from their plantations, it was very difficult to procure food, and to suffer the torments inflicted by all kinds of insects. They smeared sap drawn from the vines over their bodies and let it solidify there.[15]

Rubber-caked native workers would then transport by foot their sap quotas to their respective concessionaires. Often, in order to compel the native workers to meet their quotas, concessionaires would capture and hold whole families as ransom. This kind of cruel and rapacious activity was permitted by the colonial authorities who saw themselves as grossly understaffed and ill-prepared to regulate the chartered companies. When Brazza returned to Gabon and other territories of the French Congo in 1905 to witness the hyperbolic exploitation of the Concessionaire Regime, he found that "[t]wenty strokes of the *chicotte* were considered a handy way to keep Africans in their place."[16] At Bangui he discovered that 47 black women and children had died of starvation while in captivity as their menfolk were forced to deliver wild rubber in-kind to the local authorities:

> Slaughter, depopulation, and emigration are produced by the ruthless measures employed; the natives become crushed and terror-stricken, the vast areas of rubber-producing land are irretrievably impaired through the frantic and unscientific bleeding of the vines by droves of natives, spurred on by the fear of the consequences awaiting them if they do not bring the required amount."[17]

Deadly forced labor practices were combined with endemic pestilence to depopulate vast swaths of the Gabonese territory. The cruel concessionary practices, for example, aggravated the small pox epidemic (1886) and the deadly "sleeping sickness" which had travelled from the Atlantic ports along the coast and up river to the interior. The most persistent infection in the region, according to Catherine Coquery-Vidrovitch, was venereal disease, which caused sterility and lowered the birthrate in extensive areas of central Africa.[18] Colonial tropical medicines were inadequate to the task of pandemic tropical diseases. Influenza alone, for example, killed 10 percent of the native population.[19]

Slavery, ruthless exploitation, and deadly diseases were the key factors that historically stunted the political development of the indigenous societies of Gabon. Tyrannical foreign domination did more than emasculate and muzzle traditional state systems, it also incapacitated and devitalized traditional societies. The populations that survived four centuries of oppression in Gabon were politically crippled in such a manner that absolutist-style government came almost without resistance.

In fact, from the standpoint of military expenditures, the cost of acquiring and defending the entire French Equatorial African territory was the lowest in any of

France's overseas dependencies. For the occupation period it has been estimated that "in terms of manpower, at most 700 soldiers were killed and 1,200 to 1,500 wounded, of whom about four-fifths were Africans."[20]

THE COLONIAL ORDER

The creation of the Colony of Gabon (1906) and its subsequent incorporation into the French Equatorial African (AEF) federation (1910-1958) marked a turning point in the history of the "state" in the country. The nineteenth-century abuses of the feral concessionaires were tamed by the establishment of a twentieth-century colonial order that took on the essential characteristics of a modern state: legal order, bureaucracy, jurisdiction over a territory, monopolization of the legitimate use of force. Because it adopted these features we can consider the colonial regime, for better or for worse, as the prototype of the modern state in Gabon.

Colonialism introduced a legal order that, while racist, did manage to subject all individuals living in the colony, black or white, to the rule of law rather than the arbitrary and unregulated justice of the company agents. Colonialism also introduced the territorial boundaries that today correspond to the Gabonese state (with a few slight adjustments and alterations) as well as a bureaucratic structure and divisions and subdivisions with compulsory jurisdiction assigned to administrative officers—replacing the concessions as the most important geographic realities. Perhaps most important of all was the monopolization of the "legitimate" use of force by the colonial officers, who became the personification of the state within their respective colonial domains. As Aimé Césaire wrote in his *Discourse on Colonialism*: "[T]he great historical tragedy of Africa has been not so much that it was too late in making contact with the rest of the world, as the manner in which that contact was brought about."[21]

It is here, at the twilight of the twentieth century, that we first see an outline of institutions which are identifiable as the "state" in Gabon. The neocolonial understanding of the Gabonese state advocated by authors such as Pierre

Péan derive from this colonial state. A French-dominated state structure that was set in place during the colonial period became the structure upon which an independent Gabonese state was later manufactured. So too the seeds of personal rule described by Jackson and Rosberg were nourished in the absolutist soil of the colonial regime.

Because government in colonial Gabon was a bureaucratic type of administrative organization, political representation did not exist in anything but the most formal sense. The bureaucracy operated on the idea that the colonial executive at each level of the hierarchy ought to possess considerable autonomy of decision-making over his territorial jurisdiction. At the top of this bureaucratic pyramid of power was the governor-general in Brazzaville, who was the supreme authority personified, and the sole intermediary between the Ministry of the Colonies in Paris and the AEF federation in Africa. This attitude of impersonal-personal rule was expressed best by M. Martial Merlin, the first governor-general of AEF, when he addressed his lieutenant-governors and territorial commandants with the following maxim: "I govern, you administer."[22]

If any individual institution planted the seeds of absolutism in the political spirit of Gabon, it was the colonial officer, who combined executive and adjudicating powers and could implement executive decrees (*arrêtés*) within the broad framework of legislation (*loi*). The Gabonese colonial state did not suffer any system of separation of powers nor checks and balances.

On the territorial level, the Colony of Gabon reproduced in miniature the government-general in Brazzaville. From his capital city of Libreville the lieutenant-governor of Gabon (also known as the *chef de territoire*) ruled the colony with only the most limited regulation from above. He appointed a purely advisory *conseil des intérêts locaux* and wielded substantial power over the *commune mixte* of the capital city itself. Beneath him were seven *chefs de département* and twenty-seven *chefs de subdivision* who each ran their corresponding jurisdictions with wide latitude.

The colonial officers monopolized the "legitimate use of force" within their respective domains, but many factors such as their often brief tenure in any one post and the limited resources available to them in the harsh and underdeveloped rain forest led the officer corps to routinely appoint native "chiefs" to assist them in the performance of their duties. For example, village chiefs were appointed by their respective *chefs de département* and were held accountable for the maintenance of village roads, for public sanitation, and for the preservation of public order. In this last duty the chiefs were given the power to arrest, but not to adjudicate or punish (as they would have had in the traditional clan systems).

Village chiefs in Gabon were also responsible for the collection of the native head-tax, and were allowed to keep 10% of the take for themselves as payment for services rendered. The village chief in this process became a shadow of his respective administrative cadre, and could be "tried, suspended, or dismissed by government and by government alone."[23] As Coquery-Vidrovitch explains in her discussion of this system, the legitimacy of a village chief's traditional authority was soon undermined by his source of revenues. Previously he would have collected tribute from his clansmen in exchange for his adjudication or other traditional fuctions. But the new system put him on the colonial payroll in the form of salaries and rebates from the head tax. As this process evolved over time, "everywhere the chief became more or less absorbed into the public administration, and his territory tended to be transformed into a simple administrative subdivision."[24]

Absorbtion of traditional village chiefs into the colonial staff was made necessary by the gross understaffing of the colonial offices. The French model of "direct rule" required a high ratio of colonial civil servant to African subject. But by 1928 only 250 out of the 366 existing posts had been filled.[25] Gabon and the AEF had one of the poorer reputations in the French empire, and few young graduates of the Grandes Ecoles in Paris wanted to be posted in its malaria-ridden swamps and jungles. The fact is that often the worst of what the empire had to offer was stationed in Gabon. Its bureaucracy left behind a legacy of concentrated and centralized power in the hands of unelected administrators. Decision-making in this kind of organization was a-political, in that it followed a despotic chain of command rather than establishing group inclusive participatory processes. The paperwork alone reflected the new order. From the bottom up, administrators in the subdivisions were made to send their requests to their department heads, who in turn had to report to the territorial governor, who in turn reported to the government general in Brazzaville. Written letters, written requests, and written orders flooded the mailboxes of the colonial administration. "This procedure had to be repeated for every facet of industrial and commercial activity," complained one official. "Weeks—even months—passed before a reply was received."[26]

The ability to read and write in the French language was basic to the colonial regime. The legal order that colonialism established in Gabon was not one based on local customs and local traditions, but rather was based on French law and French culture. To the degree that a native black African could adapt himself to the language, habits, customs, and tastes of his oppressors, he was able to cooperate with and participate in the general colonial project. An official policy of *assimilation* was promoted in Gabon in order to accomplish this project. *Assimilation* can be defined as a "process by which non-French peoples were to be assumed into the body of the French nation, taught its language, and indoctrinated in its culture."[27] The process was pursued in Gabon at first by the Brothers of Saint Gabriel, who arrived in Libreville in 1900 to establish the Ecole Montfort, and a similar boys' school in 1901 at Lambaréné. These mission schools taught bright young Gabonese to read and write in French, indoctrinated them with Catholicism, and generally assimilated them into French civilization and culture. Michael Reed dates the beginning of Gabonese politics to the foundation of these mission schools, because the first and second generations of nationalist politicians were educated by them. Even when in 1905 the separation of church and state in France caused the closure of the mission schools for a twenty-year period, the requirement that all secular teaching be done in French "ensured that the emerging Gabonese leadership would be firmly francophone as well as probably francophile."[28]

The most enduring legacy of the colonial educational system was that the mission schools managed to cultivate a small Gabonese elite who, in the words of Michael Reed, proved to be "elitist, non-radical ... and almost always pro-French."[29] Successful *assimilation* in Libreville produced a generation of nationalist leaders who, even when they were anticolonial, were seldom anti-French. In fact, claims Reed, Gabon has never produced a tradition of radical anti-French politics. On the contrary, the Gabonese became among Africa's most proficient speakers of French, who styled their schools and other institutions on the French model they themselves had known. Even Gabonese opposition parties were based in France. "The first national leaders," writes Brian Weinstein "were mainly civil servants or employees of private companies who . . . had received Roman Catholic education at the Ecole Montfort."[30] Weinstein argues that the nature of their politics was shaped by their early formative experience. Jean-Hilaire Aubame and René-Paul Sousatte, for example, were educated together at Ecole Montfort, then went on to serve together in the colonial civil service in Brazzaville, before going on to become the spokesmen of opposition parties. Although associated strongly with the idea of independence and opposition in the Gabonese political spectrum, both men used their education and contacts to advance themselves within the colonial order. Aubame served as a representative to the French National Assembly and Sousatte to the Union Française in the 1940s. Neither man participated in active resistance to colonial domination, and certainly neither participated in anything even remotely resembling armed struggle.

FALSE INDEPENDENCE

Armed struggle would have been, at any rate, superfluous. For unlike other French colonies that had to struggle for their freedom (e.g., Indochine, Algérie), it was the good fortune of Gabon to have been made free without having to become so. The making of the state in Gabon was a peaceful process devoid of the violent anticolonialism known elsewhere. No revolutionary movement ever challenged the colonial regime in Libreville. No guerilla troops fought colonial forces. No terrorists took Frenchmen hostage in Gabon demanding their country's independence.

Instead, a long 16-year process of nonviolent and orderly decolonization, beginning with the Brazzaville Conference in 1944 and ending with the autonomous Republic of Gabon in 1960, deposited freedom, as it were, in the laps of the Gabonese elites. Because no revolutionary anticolonial movement emerged from the Gabonese masses, decolonization and independence became an elite undertaking. In fact, the first major reform of the colonial regime instituted after the Brazzaville Conference was the creation of territorial assemblies and admission of two African representatives to the French parliament. These republican institutions favored elite participation and, in fact, saw the election and appointment of Ecole Montfort-educated elites such as Jean-Hilaire Aubame, Rene-Paul Sousatte, Paul Indjenjet Gondjout, and Léon M'Ba (whose brother the priest was a favorite of the Brothers of Saint Gabriel).

The Brazzaville Conference, in the myth-making process that often surrounds such historical events, has generally been credited with starting the reformation of the French African empire that earned General de Gaulle the appellation "Father of Independence" for the francophone black Africans. De Gaulle made no specific promise to end colonialism, and "no mention was made at this time, at least by de Gaulle, of the idea of self determination."[31] But he did promise that France would, in his words "proceed, when the time is opportune, to make reforms in the imperial structure."[32] At the Brazzaville Conference recommendations were made for the enfranchisement of native Africans and for their representation in the legislative assemblies of the empire. "It was impossible" wrote Jean-Hilarie Aubame years later, "to continue to think according to the old colonialist conceptions."[33] For in opening the doors to the French parliament, the Brazzaville Conference introduced the idea of African representation to Gabon.

In a national referendum held in France in 1946, French voters fulfilled de Gaulle's promise by granting African *territoires d'outre-mer* limited representation in the National

Assembly, the Council of the Republic, and the Assembly of the French Union. In this new arrangement Gabon received one deputy, one senator, and one representative to the French Union. In 1956, the French parliament passed the *loi-cadre*, which was enabling legislation that authorized representative Territorial Assemblies in both the West and Equatorial African federations. By expanding the representation of Gabonese to the French parliament and to the new Territorial Assembly, France not only vented the more radical anticolonial sentiments of her African possessions, but she also socialized the Gabonese elites in the French political discourse, structuring the future debate of decolonization and independence in terms that reflected the French ideological spectrum.

The first generation of important Gabonese politicians cut their teeth on the political institutions of Paris, and despite their differences of ethnicity, shared a common experience that served to reinforce and complement their pro-French attitudes. Jean-Hilaire Aubame, for example, who served as Gabon's first representative to the French National Assembly (1946-1958), and who was the leader of the Fang-dominated political party, the Union Démocratique et Sociale Gabonaise (UDSG), became a strident advocate of multiparty parliamentary democracy of the kind he had been initiated in during his years of service under the Fourth Republic. Similarly, Paul Indjenjet Gondjout, who was a Mpongwé from the Estuary who served as Gabon's senator to the Council of the Republic (1949-1958) and who founded the Bloc Démocratique Gabonais (BDG) to oppose Aubame's Fang threat to the traditional Libreville elite, shared with his rival a common vision of legislative supremacy, and in fact joined forces with Aubame to prevent the presidential system that was imposed in the first year of political independence by Léon M'Ba. Another of this first generation was René-Paul Sousatte, who founded a Punu-Eshira mass-based political party, the Parti de l'Unité Nationale Gabonaise (PUNGA); and once again, despite the fact that he personified the southern ethnicities' opposition to the domination of northern ethnic groups (such as the Fang and the Myéné), his early formative years as Gabon's representative to the Assembly of the French Union in Paris gave him a shared politicization and pro-French attitude.[34]

The power of the business community in Gabon expanded during the postwar era into the new political arena. The opening up of political space for representation also created an opening for the business community to participate in the electoral process. Elections, as modern political science has shown, are very good vehicles for business to participate in governance. First they can provide an official role for successful businessmen to play within the framework of the state itself. For example, Luc Durand-Reville, the director of Gabon's biggest and most powerful forestry concern, Société du Haut Ogooué (SHO) was elected as a senator in the French Council of the Republic (1946-1949) and from there involved himself actively in the Gaullist RPR. Second they can provide a legal conduit for money to be channelled in politically desirable directions. For example, Durand-Reville and the Libreville Chamber of Commerce became major backers of Paul Gondjout and with the assistance of Mpongwé businessmen and the foresters helped elect Gondjout to the Council of the Republic. So powerful were the business interests in Gabon that it is not hyperbole to say that moden Gabonese politics would not make sense without understanding the forces of capital behind them.

The most extreme case of business' power was the forestry lobby under the leadership of Roland Bru. We have already discussed in general terms how influential forestry was to the colonial economy of Gabon. But the specific role of Bru in Gabonese politics has been called "the most extreme yet exemplary case of the role of French businessmen in Gabonese decision-making."[35]

Roland Bru emerged as the leader of the foresters in the late 1940s and early 1950s during the struggle over *okoumé* prices between the colonial Office des Bois and the Syndicat Forestier du Gabon. The Office des Bois had been created after the war to control the purchase and sale of lumber—especially *okoumé*—and to determine its price in accordance with the needs of the colonial metropole. In response the foresters had organized themselves into the Syndicat Forestier du Gabon to protect themselves from what they feared would be ruinous prices. As forestry was the number-one industry in Gabon, an association of this kind had obvious political ramifications. Joan Edelman Spero has described the Syndicat as "the most important business group both financially and politically" in the colony.[36] Forged in the common struggle to maintain high *okoumé* prices, the Syndicat soon became actively involved in lobbying. One of Bru's early accomplishments, after the buttressing of the *okoumé* price, was his acquisition of public money from the Fonds d'Investissement pour le Développement Economique et Social (FIDES) for the construction of several large sawmills along the banks of the Ogooué river system. The effect of this large-scale mechanization was to drive out small concessionaires and to concentrate even more power in the hands of the Syndicat. As postwar reforms opened up the possibility of limited self-rule by the colonies, Bru and the foresters began to support candidates who voiced their common concerns (e.g., Durand-Reville). With the clout of the Syndicat behind him, Bru became for all intents and purposes, a "kingmaker" in Gabon.

Bru and the foresters began to feel threatened by the rising popularity of Aubame and his UDSG. Aubame had earned a reputation as a political moderate, and was well-liked by the administration. But as the *okoumé* pricing conflict illustrates, the interests of the administration and those of business did not always converge. Bru and the foresters needed their own man in office, someone who would represent their interests in the new political arena created by the postwar imperial reforms.

They found their man in Paul Gondjout, who feared the rise of the Fang majority in Gabon, and whose conservative business roots in the Mpongwé commercial class of Libreville placed him in ideological opposition to Aubame's socialism. Gondjout had made the friendship of Durand-Reville during his Paris days, and both men had served in the Council of the Republic as senators for Gabon. As Barnes writes, "in his eyes, and those of the non-Fang estuary elite, Aubame represented a real threat to their traditional hegemony."[37] But as the ethnic group from which he came had dwindled over the years to a statistically insignificant size, Gondjout recognized that his success resided in his ability to transcend the politics of ethnicity and to build an interethnic coalition party. With the support of Roland Bru, Luc Durand-Reville, and substantial contributions from the timber industry and Mpongwé businessmen, Gondjout joined forces with the Fang politician Léon M'Ba to found the Bloc Démocratique Gabonais (BDG) in 1954 as an opposition party to Aubame's UDSG. This new party aligned itself with the Rassemblement Démocratique Africain (RDA) of Houphouët-Boigny, distinguishing it from the UDSG's affiliation with Senghor's PRA.

THE RISE TO POWER OF LEON M'BA

The rise to power of Léon M'Ba represents one of those great political come-back stories, yet it also reveals the method by which Bru and the foresters were able to groom the future leadership of the country. Léon M'Ba was born the son of a Fang village chief in Libreville, and at the age of 22 was appointed the *chef de canton* for the entire Estuary Fang. During the 1920s and 1930s M'Ba made a reputation for himself as someone willing to fight for the Fang in their land disputes with the Myéné. He also "incurred the suspicion of the administration" by becoming active in several radical organizations, including the Ligue des Droits de l'Homme and Jeunes Gabonais.[38] He published letters in Libreville newspapers and journals on behalf of illiterate clansmen to voice their grievances, and associated himself with the French communist party in his youth. Finally, his participation in the *bwiti* secret society, which was seen by the missionaries as a threat to the Catholicism from which it had drawn many of its elements, and by the colonial administration as a Fang nationalist challenge to the existing imperial order, landed him in jail in 1931 on charges of murder and ritual cannibalism.[39] He was imprisoned until 1934, then transferred to Oubangui-Chari, where he was employed by the AEF civil service, and was later able to distinguish himself during the Free French Resistance movement (1940-1944). In reward for his loyalty, M'Ba was allowed to return from exile in 1946. He was 44 years old and, matured by his years of active service, decided to establish himself at a modest but respectable post in the British firm John Holt. From there he befriended Libreville Chamber of Commerce Officials.

His electoral career started in the late 1940s, when he took over an all-Fang political organization and founded the Comité Mixte Gabonais (CMG). At the same time he became the president of the Groupe d'Etudes Communistes (GEC), which was an organization sponsored by French administrators sent overseas during the Popular Front government in France. This organization did not require membership in the Communist party and M'Ba himself was far too practical to adhere to any ideology. Instead his work with several representatives of other ethnic groups in the GEC helped him to expand his political base outside the narrow Fang electorate. In 1951 he broke with Aubame and directly challenged him in the Territorial Assembly elections. Although unsuccessful at his bid for the office, M'Ba did manage to attract the attention of prominent French expatriates living in Libreville. In order to include them in his program, M'Ba changed the name of the CMG to the Comité Mixte Franco-Gabonais. In his campaign literature, M'Ba emphasized his willingness to work with other ethnic groups. "Whatever your color, whatever your race or tribe," he exorted in a campaign pamplet for the legislative elections of June 17th, 1951, you should vote for Léon M'Ba because the "personal relations he had maintained since 1922 with the diverse ethnic groups of Gabon" made him the candidate most "capable of forging a union of all Africans and Metropolitans" of the colony.[40]

Undiscouraged by his electoral defeat in 1951, M'Ba won many admirers for his effort to build an inter-ethnic coalition that could compete against the increasingly influential, Fang-dominated UDSG. M'Ba also developed a "populist" style that contrasted with that of his counterparts, Aubame and Gondjout; for unlike these men he had missed the opportunity to be politically initiated in the genteel institutions of Paris. His early radical activities and exile forced him to take the hard road to power. There was talk of his supposed connection with strange occult rites which gave him *evus* or personal power, and earned him the reputation of a witch and a man to be feared. But he was equally well known as a "captivating person" and a "man of the people" who was "earthy, appealing, passionate."[41] Using these charismatic talents, M'Ba finally succeeded in winning a seat as a deputy in the Territorial Assembly for the Estuary in 1952. Like the prodigal son, twenty years after his imprisonment by the colonial authorities, he had come back to become a part of the colonial establishment.

In 1954 he was invited to join Gondjout and the BDG, and that year he became the party's Secretary. The Gabonese elites and Frenchmen worked closely together in the early political parties of Gabon and were dependent on alliances with different French groupings within the colony: "Aubame's UDSG on the French administration; Gondjout's and M'Ba's BDG on French lumbermen, especially Roland Bru."[42] For the next decade M'Ba would remain aligned with Bru and the foresters on his way to the top of the political ladder.

Only with the financial support of the foresters and the business community was M'Ba able to win the 1956 mayoral race for the city of Libreville. Following the traditional French practice of holding multiple positions—*cumul des positions*—M'Ba now served as both mayor and deputy for the Libreville constituency. But the source of his financial backing came to be common knowledge and undermined his credibility with much of the native electorate. In the 1957 elections for the Territorial Assembly, when his BDG attempted to once again unseat Aubame, this reputation as the "forester's man" worked against M'Ba, who found himself unable to reduce his rival's popularity with the people. The UDSG won a majority of the popular votes in this election. However, in what has come to be considered the definitive study on the development of political parties in AEF, John Ballard has shown how Bru and the foresters effectively paid for the victory of M'Ba and the BDG by providing funds for their candidates. Bypassing the popular mandate of the people, the foresters bribed several UDSG deputies to switch affiliation to the BDG. Additionally, the legislative districts were gerrymandered to benefit BDG candidates, with the result being a popular majority for the UDSG, but a BDG-dominated Territorial Assembly.[43]

In this manner, M'Ba became the vice president of the Government Council (under the presidency of the colonial lieutenant governor). It is only a slight exaggeration to say that his position was bought and paid for, and that from it he would act as a puppet of Bru and the foresters.

The year 1958 was a watershed in the history of the state in Gabon, for it was the year that Gabon became an autonomous republic separated forevermore from Brazzaville and the AEF federation. The events leading up to the fall of the French Fourth Republic in 1958, such as the military defeat of French troops at Dien Bien Phu (1954) and the bloody Algerian war of independence (1954-1962) and the Army's revolt in May of 1958, culminated in the return of General de Gaulle to power, the drafting of a new constitution for the Fifth Republic, and the creation of a French Community of nations premised on the British Commonwealth. De Gaulle offered a referendum to France's African colonies in which they could either choose to remain within the new Community as autonomous republics or secede from France to outright independence. On a tour of French Africa in August of 1958 de Gaulle warned the colonies that a "no" vote on the referendum (rejecting the Community) would result in the immediate withdrawal of all French aid. This threat was very persuasive in Gabon, where FIDES alone had spent around 110 billion CFA francs between 1946 and 1958, and where the public finance regime was almost entirely aid based.[44] By an overwhelming majority, Gabonese voters accepted de Gaulle's proposal (190,334 to 15,244) and remained within the French Community.[45] Neither Aubame nor M'Ba expressed any desire for full independence, but rather feared that total freedom would be ruinous for Gabon at that time. Also both politicians were cognizant of the advantages being offered by their new autonomy from Brazzaville and the AEF. For the next two years (1958-1960) Gabon remained an autonomous overseas territory within the French Community nominally governed by M'Ba, the BDG, and the Territorial Assembly.

On August 17th, 1960, during a wave of decolonization that swept across the map of francophonic black Africa, the colony of Gabon gained its independence from France and became the Republic of Gabon. Léon M'Ba became the country's first prime minister and "real" head of government, albeit by virtue of the parliamentary majority that had been purchased for him by Bru and the foresters. And since the Republic of Gabon was a classical parliamentary system, in which the head of government was also effectively the head of state (being that no traditional monarch existed who could claim to represent all of Gabon) it would have been safe to assume at that time that M'Ba was also the head of state.

But the state had not been made in such a way that M'Ba could head it. This was the case not only because the cooperation accords and general terms of independence preserved France's dominance, but also because the parliamentary system which Gabon had inherited from her former colonial metropole did not provide for a strong executive. For the first year of political independence M'Ba was made to govern a coalition government that included not only his political allies such as Gondjout and the BDG, but also included his political rivals such as Aubame and the UDSG, and Sousatte and PUNGA. The same institutional weakness which had conditioned the collapse of the French Fourth Republic—*immobilisme* of multiparty regimes—threatened to undermine the stability of its Pygmalion system, the Republic of Gabon. Critics voiced their displeasure with M'Ba and his reliance on French administrators. They strongly objected to his appointment of Roland Bru, symbol of the old colonial order, to his ministerial cabinet. From that moment on, "M'Ba moved to eliminate his rivals and all effective opposition to his personal rule."[46]

In 1961 M'Ba called for a constitutional convention that would change the parliamentary system into a presidential one like de Gaulle's French Fifth Republic. This caused his longtime ally Gondjout to organize a motion of censure against M'Ba in the assembly. M'Ba accused Gondjout of plotting against him, and had the elder statesman arrested. Gondjout was convicted of the charges of attempted coup and sentenced to two years in prison. Sousatte, who also opposed M'Ba's effort to create a presidential system, was arrested and convicted and sentenced. This left only

Aubame to challenge M'Ba and his presidential aspirations.

Following the maxim to keep his friends close and his enemies closer, when the time came for these men's release, M'Ba appointed them to silent positions in his government. He appointed Gondjout to the Economic and Social Council upon his release from prison. He also appointed Sousatte as the Minister of Agriculture. Both of these positions were symbolic at best, and effectively removed the two politicians from any possible threatening position of real power in the country. M'Ba also tried to neutralize Aubame by naming him head of the Supreme Court, and then passing legislation prohibiting the *cumul des positions* by which Aubame retained both his legislative seat in the assembly and his new position on the court. M'Ba expected Aubame to keep the Supreme Court seat, given that it was the more prestigious of the two positions he held. But Aubame chose to relinquish his court seat and retain his legislative one, thereby remaining a thorn in the side of M'Ba and his presidential ambitions. Unable to silence Aubame or remove him from the assembly, M'Ba proceeded to ignore his rival, and pressured the BDG to pass his legislation creating the office of the president. The legislature did this by passing a constitutional amendment, and in 1963 M'Ba achieved his long-sought-after wish. As president of the Republic of Gabon in 1964, M'Ba dissolved the assembly and called for elections creating a one-party state.

THE FOCCART NETWORK

While Léon M'Ba attempted to consolidate as much power as he could in his own hands, and those of his party, another less conspicuous effort was being undertaken by de Gaulle's right-hand man for African affairs, M. Jacques Foccart. The story of Jacques Foccart and his role in shaping Franco-African relations in the postindependence years is essential to a proper understanding of the state in Gabon. By an order of December 19th, 1958, de Gaulle created a Secretary General for the Community which would come to be the most influential post in the Fifth Republic on all matters African. How Foccart came to be so influential deserves a moment's digression.

Born in 1913 at Ambrières-le-Grand in Normandy, Jacques Foccart was raised by a family of farmers and merchants of relatively modest middle class origins. The Foccarts ran a family business on the island of Guadeloupe, where they cultivated the attitudes common to the French *colons* during the Third Republic. Jacques spent his youth developing his business acumen and by the age of 22 prepared himself for a career in the import-export trade. But his entrepreneurial aspirations were interrupted by the outbreak of the war with Germany. In 1940 Foccart joined the French resistance under the assumed name of "Binot" and received training in the paracommando units that operated behind enemy lines during the allied invasion of 1944. During the final stages of the war Foccart rose to the rank of lieutenant colonel. This early paramilitary experience established not only Foccart's impeccable wartime credentials as a hero and a patriot, but it also exposed him to the secrecy and subterfuge of his future organizational activities.[47]

During the war Foccart made his connections with Gaullist soldiers and spies that later formed the basis of his secret and powerful network. The Bureau Central de Renseignement et d'Action (BCRA) in London became one source of connections. It had been founded in 1942 to remove the Nazis from France and to prepare for the installation of General de Gaulle to power.[48] Foccart's service with the wartime paracommando outfit DGER which would later become the Service de Documentation Extérieure et de Contre-Espionnage (SDECE)—the French equivalent of the CIA—equally prepared him for his role as the mastermind of a Gaullist paramilitary network.[49] Comrades from his wartime adventures became loyal members of the pool of readily available thugs and assassins (*la piscine*) used by Foccart when needed to effectuate his personal foreign policy in Africa. In 1946 a special "Action Service" (*service "Action"*) of the SDECE was created to, in the words of Pierre Péan, "instruct men to kill, steal, and commit the most illegal coups in the name of its own survival."[50]

Foccart's public life after the war centered around his involvement with de Gaulle's Rassemblement du Peuple Français (RPF), which he had helped to launch in 1947 and in which he became de Gaulle's number-two man. The RPF was considered the most imperial of the postwar political parties in France, and within this framework Foccart earned the reputation as *le penseur du "gaullisme colonial"* who wanted to recover the prewar status quo ante.[51] In a series of articles written by Foccart in the *Lettre à l'Union Française*, Foccart warned his countrymen against the dangers of communism, the threat of the USSR, and the need for France to retain her prewar imperial order. In one article he wrote that "the future of the Rassemblement could depend on the influence that we have acquired in the overseas departments and territories."[52] To this end, he befriended Félix Houphouët-Boigny of the Ivory Coast and aligned the RPF with the RDA, and was instrumental in the Ivorian leader's break with the French Communist Party (PCF).

The other side of his activities, those outside the public limelight, were best described as "occult."[53] Foccart became a close friend of de Gaulle during the postwar years and worked for the General's return to power. According to Péan, Foccart was actively involved in the destabilization of the Fourth Republic, described by one Gaullist cabinet member years later as "the key man in the conspiracies of May 13th."[54] Foccart had slowly gained control of the *service d'Action* of the SDECE and had infiltrated the secret service with likeminded partisans of de Gaulle. Through his import-export business, SAFIEX, Foccart was able to network private spies and agents throughout French Africa with his SDECE and RPF comrades, building the infamous *réseau Foccart.*[55]

By December of 1958, when de Gaulle created the office of the Secretary General for the Community, Foccart was brought into the Elysée as a technical adviser on African affairs and as a liaison between the president and the SDECE. Then de Gaulle appointed Foccart Secretary General and made him responsible for maintaining contact with African heads of state. In 1960 the title was changed to Secretary General for the Community and for African and Malgasy Affairs. Theoretically the Ministry of Foreign Affairs (*Quai d'Orsay*) should have been responsible for the conduct of France's foreign policy toward the African and Malgasy states, but in reality this responsibility was monopolized by Foccart. He and his staff at the Elysée may have invited ministers from the Quai d'Orsay to come to his offices for the weekly meetings he held to coordinate inter-agency activities on African affairs. But fundamentally, it was Foccart who dictated France's African policy. Foccart and his staff prepared all presidential decision papers on issues that involved Africa during both de Gaulle's and Pompidou's presidencies. Foccart also organized all visits by African dignitaries and heads of state to France, as well as all visits to Africa by the French presidents. "Other participants at the meetings were painfully aware that the secretary general met daily with General de Gaulle" writes McNamara. And because no other minister had the same access, "only the most intrepid or foolish bureaucrat dared question Foccart's authority to speak ex cathedra for de Gaulle or Pompidou."[56]

Foccart's extraordinary influence within the French government on the matter of African affairs was conditioned by the fact that the Ministry of Foreign Affairs had almost no tradition of foreign policy for the African states before the 1960s,[57] as well as Foccart's intimate personal ties with de Gaulle dating from his days in the resistance and the RPF. But perhaps the most important of all was his influence outside the framework of government and law.

It was arguably his personal relations with people from all walks of life, his ability to create network where none existed, that made Foccart so influential. For example, although his company SAFIEX was created by Foccart for the purposes of conducting a profitable import-export trade, it also "specialized in providing a cover for information gathering and undercover activities of the French special services."[58] According to Spero, many former agents of the BCRA during the wartime years later became employees on the SAFIEX payroll after the war. As McNamara observes, thanks to this unofficial network, "little went on in the ports and airports of Africa that Foccart was not quickly made aware of."[59]

Another example of his personalized style was his close relationship with many of the presidents of the newly independent African states. For example, Foccart cultivated a longstanding friendship with Ivorian president Félix Houphouët-Boigny, who in turn was on close terms with president Léon M'Ba. It was by way of Houphouet-Boigny that Foccart and M'Ba became personal friends. It was reputed that M'Ba could call Foccart personally on the phone and receive an audience with him at a moment's notice. While access to other countries would have had to have been made through formal diplomatic channels, the personalized style of Foccart's network gave him more influence than his position as secretary general would have technically provided.

But ultimately it was his longstanding connections with the SDECE officers and with the "*piscine*" that gave him the muscle he needed to project an aura of both power and mystery that have made him a legend in his own time. For example, colonel Maurice Robert, who headed the African service of the SDECE during 1959-1973, became one of Foccart's closest friends. Then there was the *barbouze* from the Algerian white resistance movement, Pierre Debizet, who became a leader of the *Service d'Action Civique* (SAC) and hired gun of Foccart. Yet another of the future key figures in the Gabonese political equation, Guy Ponsaille, also served in the *service d'Afrique* of the SDECE before joining the Foccart network. Men such as these first earned their reputations as capable soldiers of France, then as willing killers of the *réseau Foccart*, then as respectable officials of the Gabonese state.

It is not surprising in this regard to discover that the mercenary Bob Denard was a regular agent of the *réseau*, engaged by Foccart for, among other activities, the assassination of Gabonese opposition leader Germain M'Ba in the early 1970s. Each of these men would eventually find a place within the Gabonese administration. Even Bob Denard, in fact, served as a "counsellor" to the Gabonese president; and Debizet as the head of the presidential guard (GP).

Cooperation accords were signed by Gabon with France upon the attainment of political sovereignty and became the basis of what evolved into a special relationship between the two countries. The cooperation accords made the state in Gabon a quasi-independent legal entity.

They secured the special monetary relationship of the franc zone, the privileged trading relationship between France and her *chasse gardée*, the special French rights to Gabon's strategic minerals such as uranium for de Gaulle's nuclear force, and military arrangements by which French troops would intervene when called upon by the Gabonese president to do so. (This last point would take on pointed significance during the attempted coup of 1964.)

Both the creation of the *réseau Foccart* and the signing of the cooperation accords seriously compromised the autonomy of the state in Gabon. If Jackson and Rosberg insist on highlighting the "personal" features of political rule in Gabon, then how can they reconcile their analysis with the existence of a secret government, occult, hidden, and behind the scenes? What is the state in Gabon? On the surface it would appear to be the government of that country, the postcolonial state institutions invented by France as part of a package deal concocted at the birth of the Fifth Republic to justify and perpetuate French rule in black Africa. But if the ostensible government of Gabon is in reality controlled and manipulated by the shadow government of Foccart and the Ministry of Cooperation in Paris, as suggested by Péan, then what we have is a hybrid, internationalized state. The political order in Gabon, part African, part French, gives the appearance of being what it claims to be (an independent African state), but hides what it really is—a French dependency.

The attempted coup of the M'Ba regime in February of 1964 is a case study in the necolonialism of the so-called "mutual" defense accords. On February 17th, 1964, a clique of Gabonese military officers led by lieutenants Mombo and Essone overthrew the government of president M'Ba and held him briefly as their prisoner. They called upon Jean-Hilaire Aubame to assume the responsibilities of head of the interim revolutionary government, and made pronouncements over the state broadcasting media explaining their actions.

The cause of the coup clearly involved president M'Ba's blatant efforts to monopolize all political power in his own hands. In 1963 he had dissolved the National Assembly on the pretence that an austerity plan would necessitate the reduction of a number of seats in the legislature. M'Ba called for new elections, in which the opposition refused to participate, and which were designed to introduce his party as the sole legal party of the state.

The revolutionary officers, Mombo and Essone, called on France to remain neutral in this affair, and not to interfere with the internal affairs of Gabon. But the next day, February 18th, Foccart met with his closest conspirators—Maurice Robert, Pierre Guillaumat, Guy Ponsaille, RenéJourniac, and Claude Terraroz—and decided to intervene on M'Ba's behalf. The official terms of the military accord required that the Gabonese executive would have to call upon France to intervene before France could do so. Foccart later claimed that Terraroz, the Gabonese ambassador to France, had been given such an order by M'Ba. But subsequent accounts indicate that this would have been impossible, given the fact that president M'Ba was a prisoner in the hands of the coup plotters. And vice president Yembit was apparently in a car with U.S. ambassador Charles Darlington and therefore could not have activated the defense accords on the president's behalf. The defense accords were therefore activated by France without proper authorization by Gabon, or in effect, activated unilaterally by France.[60]

Foccart sent Robert and Ponsaille to join the 7th RPIMA in Dakar, which had been put on alert a few hours earlier, and from there to land in Libreville. Foccart had given them the strict orders to "normalize" the situation by February 19th, or the next day at latest. With the help of regular French troops, Robert and Ponsaille elicited several members of the *piscine* to join them for the dirty work, and in the words of Péan: "A Libreville, une heure avant le lever du soleil, le 19 février, les barbouzes françaises dégagent la piste de l'aéroport des obstacles."[61] Within hours the revolutionary government under Aubame surrendered, and after a few small skirmishes resulting in around fifteen dead Gabonese officers and two dead French paratroopers, the coup was effectively terminated. M'Ba was released unharmed.

"In the following months there was an orgy of arrests and beatings," Darlington informs us. Squads of thugs known as "*les gorilles*" circulated through Libreville arresting anyone suspected of opposing M'Ba."[62] The president himself became reclusive, shut up in his presidential palace during this time. There he was surrounded by French advisers and protected by French troops. After four years of living under the illusion of being an independent African state, president M'Ba had a rude awakening to the cold, harsh political reality that he was little more than a puppet of French design. Imprisoned in his presidential palace, surrounded by Frenchmen of dubious moral character, he must have known that his thoughts would have to be *their* thoughts, his words *their* words, his presidency an instrument of *their* will.

It was at this time that cancer began to take root in his body. Darlington described him as a "very isolated man" with whom even the vice president could not get an audience. Foccart's agents turned M'Ba against Darlington and the Americans, accusing Peace Corps officers of actually being CIA functionaries, and stirring up a general anti-American sentiment. According to Péan, Ponsaille and Robert "were never far" from M'Ba, and counselled him to conduct mass arrests and imprisonment of 150 opponents of his regime, including Aubame, who was held responsible for the coup and would remain in closed quarters, then house arrest, for the remainder of M'Ba's natural life.[63] Darlington writes:

> For all this sordid work France bears a large share of responsibility. France protected Leon M'Ba with French soldiers, not just for a few days but indefinitely, and when this book went to press in the summer of 1967 they were still in Gabon. Behind them he was safe, safe not only to govern, but also to give free reign to his passions. Frenchmen in the Ministry of Information helped him to keep the public misinformed. French officers in charge of the gendarmery helped him to carry out his wholesale political arrests and brutalities.[64]

Foccart's countercoup altered forevermore the nature of the state in Gabon. It did this in part by installing a permanent French military contingent just outside the capital city, which served to discourage any further outbursts of popular discontent in the country. It also did this by installing a new set of political actors, both French and African. The French military force became in effect the Gabonese army, air force, navy, and marines.

But even more important were the French civilians who were sent by Foccart to take over the government and administration of this tiny excolonial enclave. Many of these civilians were previously military men who served after the war in the *piscine*. Ponsaille, for example, was appointed the head of Elf's oil operations in the Port-Gentil region. Robert became a frequent visitor to Gabon as Foccart's go-between, and eventually found employment in Elf-Gabon.

The big shift came in the enclave industries, where Roland Bru and the foresters were replaced by Ponsaille and the oilmen. Foccart blamed the 1964 coup on what he perceived to be Bru's incapacity to control radical elements of Gabonese society. In Foccart's eyes his wartime allies were better trained to deal with threats to France's vital interests in the country. As Péan writes: "*Elf bouscule les forestiers. Apres le putsch de février, Bru ne retrouvera jamais plus son poids d'antan.*"[65] After 1964 Gabon became ruled by a handful of Foccart's men who came to be known as the "*Clan des Gabonais.*" These men surrounded the old and now shaken Léon M'Ba, effectively isolating him from the people over whom he ruled. They also replaced M'Ba's former African allies. The most important staff change of this kind was the introduction of Omar Bongo to the M'Ba presidency and his postcoup rise to power.

OMAR BONGO AND THE CLAN DES GABONAIS

Omar Bongo came to power from within the system, quietly yet ambitiously working his way from one office to the next, making his reputation where it mattered with the Frenchmen who controlled the country. His story contrasts starkly with the flamboyant and populist political life of Léon M'Ba. Born "Albert-Bernard" Bongo on December 30th, 1935, at the village of Lewai in the district of Lekoni in the Haut-Ogooué region of southeastern

Gabon, he was the benjamin of a family of nine children who lost his father when he was eight years old.[66] Unlike Léon M'Ba who came from the family of a traditional village chief, Albert-Bernard was raised by an uncle without the promise of a brighter future. Unlike Léon M'Ba, who came from the largest ethnic group in Gabon (the Fang), Albert-Bernard came from one of the smallest (the Téké, or Batéké). Unlike Léon M'Ba, who came from the most important city in Gabon, its capital, Libreville, Albert-Bernard came from a tiny village in one of the remotest and least-developed regions of the country—so remote, in fact, that it was not officially part of Gabon until 1946! Unlike Léon M'Ba, who was educated by the brothers at the elite Ecole Montfort, the young Albert-Bernard was sent to the official Bakongo school at Brazzaville, and then to technical lycée. Unlike Léon M'Ba, who was appointed canton chief by the colonial administrators, the young Albert-Bernard was appointed a clerk in the colonial postal service in Brazzaville. In 1958 Bongo joined the French Army Air Corps and became a second lieutenant stationed successively at Brazzaville, Bangui, and N'djamena (Fort-Lamy). The army, he recalls in his autobiographical preface to one of several books he authored about himself, was "*pour la jeunesse, une bonne école, une école où se forgent des amitiés.*"[67]

The military service did prove to be a good school for Bongo, as it had been for many other members of the *réseau Foccart*, as both a profession wherein he made friends with members of the *piscine*, and as a place wherein he established his credentials as a loyal Gaullist. He left the Army with the rank of a lieutenant, returned to Gabon, and found work at first in the Ministry of Foreign Affairs (ironically under Aubame) and then in the cabinet of president M'Ba. In 1962 Bongo was made the director of the cabinet, and "[r]apidly made himself familiar with all the secrets of Gabon."[68]

After the attempted coup the members of the *Clan des Gabonais* found in Bongo a man who they could trust to understand the special relationship that existed between their two countries. "Everything came before him," writes Péan. "He discovered that there were secret funds, for the deposit box was in his office."[69] He cooperated willingly with Foccart and as a result became the beneficiary of Foccart's largesse. During the first months after the attempted coup, when president M'Ba was isolated in his palace with Foccart's representatives, it was Bongo who administered the financial requisitions for creation and housing of the presidential guard (GP). And as the old president became increasingly reclusive and retiring, it was Bongo who gradually picked up one government portfolio after another: Minister of Information, Minister of National Defense, Minister of Tourism, Commissioner of the State Security Court, etc. In 1967 he replaced Yembit as Vice President of the Republic.

Bongo's selection for this last office is illustrative of the depth of Foccart's influence over Gabonese political affairs. The story, as recounted in the memoirs of then ambassador to Gabon, Maurice Delauney, begins in 1966 with the sudden illness of Léon M'Ba.[70] According to Delauney, the old man had never quite recovered his former joy of living after the attempted coup, and had been increasingly sick and tired during his last years in office. In 1966 he flew to Paris for a physical and was diagnosed with cancer. As Reed suggests, "[t]he French could not afford to lose the francophile M'Ba. As he lay dying of cancer in a French hospital, they groomed his successor."[71]

While M'Ba underwent treatment, Bongo took over practically all the responsibilities and duties of the president. Delauney, who was a veteran of the Resistance and an eight year colonial commandant in Cameroun before being named by Foccart as ambassador to Gabon, preferred Bongo over then vice-president Paul-Marie Yembit, whom he called in one memo "almost illiterate and totally ignorant of the way you rule a developing country."[72] In this memo he praised Bongo's "indispensable collaboration" and suggested his promotion to the vice presidency. M'Ba however did not want to replace Yembit, nor did he want to change the constitution to an American-style presidency in which the vice-president (which does not exist in the French-style presidential system) succeeds the president.

According to Delauney's memoirs, he and Foccart hounded the dying M'Ba at his hospital bed, pressing him to accept both the replacement of Yembit with Bongo, and the succession of the vice-presidency. Péan writes: "*Delauney abuse. Il va fatiguer le vieil homme usé ... Finalement, épuisé, Léon M'Ba accepte la comédie de nouvelles élections avec Bongo comme colistier.*"[73]

From his deathbed in Hospital Bernard in Paris, Léon M'Ba ran for the presidency of Gabon one last time, with his running mate (colistier) the young Albert-Bernard Bongo. The elections were, of course, staged to lend credence to Delauney's selection of Bongo. And if nothing else the unbelievably high turnout (over 99%) for this scrutin should indicate that it was as fraudulent as the offices for which it was held. Few things in nature or society occur in ratios of 99%. But this is the manner in which Albert-Bernard Bongo became the president of Gabon, an office he has held for twenty-nine years (1967-1996).

THE RENTING OF GABON

National unity and reconciliation were the primary goals that president Bongo advertized during his first years in office. "All my actions are aimed at a single goal," he wrote in 1968, "to resolder the unity of the nation."[74] In order to achieve economic development, he argued, he would have to obtain the willing participation of all his "Gabonese brothers in the realization of the grand design of emancipation."[75] To show his sincerity, president Bongo ordered the release of his longtime opponent, Jean-Hilaire Aubame, and issued a general amnesty for the remaining coup plotters. The release bespoke president Bongo's willingness to forget the past and start anew. But it was also a practical manoeuvre of political consolidation of power, drawing the nominal leader of the opposition into the new regime.

Reconciliation with his opponents did not mean that president Bongo was opening the doors to public criticism and political dissent. On the contrary, on March 12th, 1968, he announced the complete dissolution of all existing political parties in Gabon and the creation of a new, single-party system of government. The Parti Démocratique Gabonais (PDG) became the sole legal party of the regime, headed by president Bongo himself, who personally assumed the role of party secretary. All dissent was henceforth to be channelled through the appropriate party channels, which had the predictable effect of screening out the most defiant and insulting critiques.

President Bongo justified the creation of the one-party state in the name of national unity. The old BDG, UDSG, and PUNGA parties "had the strong tendency to identify with ethnic groups, transforming these organizations into defenders of certain special interests."[76] Multiparty politics was inappropriate to Gabonese realities, he argued, because it fostered ethnic divisiveness rather than national unity. Only through the vehicle of his PDG could the Gabonese overcome their ethnic factionalism and achieve the *volonté consciente d'unité* that was, for him, the very essence of a nation. In an interview granted to Marc Aicardi de Saint-Paul for *Afrique à la Une*, he explained his position:

> The briefest glance reveals the differences between a political party in Africa and one in Europe. In Europe, political parties meet face to face over ideological and political issues. In Africa, particularly in Gabon, experience has proved that political hostility rapidly becomes tribal. In Africa, the individual does not vote for an idea or an issue but for a man, the man from his province ... [PP] To sum up, if we were to return to the multiparty system while the national conscience is subordinate to the tribal or regional, I do not see how the system would work.[77]

Under the new system, patronage and rewards associated with one's ethnicity would come to an end and, as Barnes put it, "[t]he emphasis was now on being Gabonese."[78] This all sounds well and good in speeches and rhetoric, but the fact remains that through the creation of his PDG president Bongo effectively eliminated the possibility of a competitive electoral process that could have introduced a democratic system. "By eliminating all formalized opposition and legitimizing only one political organization, Bongo created a monopoly over the decision-making process."[79]

The organization of the PDG is ubiquitous, spreading through the framework of the government and administration like a tree-climbing vine. The entanglement of party and state occurs at every level and branch. At the top is the president, who is also the general secretary of the party. The party congress is composed of governmental administrators and representatives from the Gabonese national assembly. There are party committees at each level of the national administration, from provincial sections of the PDG to department, district, and commune committees. Even the trade union movement is consumed by the party, having been incorporated as a part of the PDG in 1973 as the Confédération Syndicale Gabonaise (COSYGA). So omnipresent is the party that it appoints all candidates for office to the national assembly, to the economic and social councils, and all the way down to the local assemblies. The PDG directs political action across the country. "To hold a position at the centre of government it is necessary to belong to the Party."[80]

The consolidation of the multiparty system into a single-party system provided president Bongo with a degree of political security inasmuch as his continued incumbency was assured. He was re-elected president of Gabon on February 25th, 1973, with 99.59 percent of the vote, and again on November 9th, 1986, with 99.97 percent of the vote. The existence of another political party might have reduced his margin of victory somewhat. In a single-party regime the nomination process replaces the electoral arena as the forum in which leadership selection occurs. As *président fondateur* of the PDG, Omar Bongo was assured the party's nomination for presidency of the republic.

To his distinct political advantage, and that of his regime, the Gabonese economy enjoyed a dramatic upturn around the same time that he performed these consolidating reforms. "Bongo's policy of amnesty and reconciliation," writes Barnes, "relied extensively on Gabon's growing wealth."[81] The late 1960s and early 1970s, as we have discussed in previous chapters, were periods of rapid growth in the petroleum and mining industries. Consequently, the president could handsomely reward those who became (or remained) loyal to his personal rule. Barnes claims that "[h]e was particularly generous to those who expressed their personal and political allegiance to his regime."[82] At the same time he did not have to perform the onerous collection of taxes that would have given credence to appeals for political representation. By the 1970s petroleum rents alone were providing the Gabonese state with as high as 90 percent of its public revenue. The direct personal income tax, by way of comparison, generated very little, as did all of the *impôts sur le revenu des personnes physiques* identified under Title II of the Gabonese tax code. This was due in part to the fact that outside the major extractive industries (i.e., oil, manganese, uranium, etc.) there was little cash income to tax in the domestic Gabonese economy. For example, an International Labour Office (ILO) survey reported in 1970 that less than 20 percent of the workforce in Gabon was employed in any of the wage-earning industries such as mining, manufacturing, construction, commerce, transportation, communications, services, utilities, etc.[83] Also, many wage-earning positions were held by expatriates, who under Section 2 of the Title II personal income tax sections of the Gabonese code were exempted from having to pay taxes on their earnings within the country.[84]

So important was the petroleum industry to president Bongo's political allocation programs that he helped personally fund a private army for the protection of the oil enclaves. The Société Générale de Sécurité (SGS) is a paramilitary force trained and equipped by Maurice Robert and Bob Denard with the assigned task of protecting the buildings and installations of Elf-Gabon. "The S.G.S. is a living symbol of the convergence of interests which one finds shoulder-to-shoulder in Gabon" writes Péan: the security forces, the oilmen, the key members of the Clan des Gabonais, and president Bongo himself.[85] According to Péan, a holding company of president Bongo called C.I.P.H.A., which is based in Luxembourg, owns a controlling 51% share of S.G.S.[86] The president also protects Elf-Gabon and the other producing oil companies in his petroleum enclaves with a *régime spécial* in the Gabonese tax code (Title III, chapter iv, section 1) for liquid hydrocarbons and natural gas sales tax.[87]

So important was president Bongo to the uninhibited

operation of the petroleum and mining companies in his country that key members of the *réseau* (Robert and Delauney) helped personally to create an elite fighting force that would protect him. The *Garde Présidentielle* (GP) is a 1,500-man army staffed primarily by Batéké soldiers, drawn from president Bongo's native Haut-Ogooué region, and commanded by French and Moroccan officers drawn principally from the *piscine*. This composition assures a force loyal to both the Gabonese president and the Frenchmen who help keep him in power. The GP received advanced military equipment and training and superior pay and perquisites when compared to the regular army. "*Globalement*," writes Péan, "*la G.P. inspire une sainte terreur à la population.*"[88]

Oil rent has also paid for president Bongo's technologically sophisticated state communications apparatus, including not only the usual national daily newspaper, but also an expensive radio and television network. The state monopoly on telecommunications media is, as you can imagine, an effective vehicle for propoganda and hegemonic control of opinion in the major urban areas where televisions are more likely to be owned. *Radio-Télévision Gabon* (RTG) broadcasts out of the nation's capital to an audience that has been estimated by the United Nations to include around 20 sets per 1,000 inhabitants.[89] In the late 1980s, installation of several cable television networks (*Antenne 2, Télé-Afrique*) have provided restricted service to a small percentage of Gabonese homes. The state radio station, Radio Gabon ("The Voice of Renovation") also broadcasts from the nation's capital and is the official voice of government in the country. Before 1990 the only daily newspaper published in Gabon was *l'Union*, owned and operation by the government and predictably loyal to the president. His monopoly on communication of political ideas has resulted in some unusual practices for the broadcasting of ideas critical to the regime. A column entitled "Mayaka" is regularly published in *l'Union* by ostensibly anonymous critics of the government. But, apparently, president Bongo himself is a frequent contributor and uses the column to criticize and thereby pressure members of his own government. A twice-weekly television program called "*Les dossiers de la télévision gabonaise*" similarly places important ministerial figures in the public spotlight where they are asked to respond to callers, including Bongo.[90]

Secret societies—in particular *ndjobi* and freemasonry—have been another, more mysterious source of president Bongo's consolidation of power in the country. Like his predecessor, Léon M'Ba, who acquired charismatic authority through his participation in the *bwiti* cult among the Fang, so too Omar Bongo has acquired a personal mystique through his ritual leadership of the *ndjobi* cult. Whether or not these men have actually possessed magical power, or *evus*, is beside the point. What does matter is that people have been willing to "believe" in and thereby legitimate such *personal* power. According to Péan, any Gabonese who is named as a minister in the government must first travel to Franceville, in the president's natal region of Haut-Ogooué, to be initiated in the *ndjobi* cult. Although shrouded in secrecy, one aspect of the ritual initiation that has been unveiled is for president Bongo to wash his feet in a basin of water and then have the minister-delegate drink its contents. "The initiated is then held to the utmost secrecy" writes Péan, "the breaking of which will be punishable by death."[91]

French freemasonry, the second circle into which any political aspirant must enter to gain membership in the ruling clan of Gabon, provides president Bongo with personal connections among the French. There are two lodges in Libreville—"Equatorial" and "Dialogue"—and the president is a grandmaster of the second while being a leader in the first. His participation in *franc-maçonnerie* gives him a unique rapport with the whites who become his "brothers" according to the rituals of this ancient fraternity. Moreover, his leadership role in the society "transforms his initial rapport of dependence into an ambiguous rapport where he holds an advantage."[92]

The methods used by president Bongo to deal with political opponents reveal him for the true Machiavellian he is, who knows "how to make a nice use of the beast and the man."[93] In 1971 president Bongo issued a *laisser-passer* (no. 318/71) to known mercenary Bob Denard, who on the evening of September 18th assassinated opposition leader Germain M'Ba and his wife in their car while they were returning home from the movies.[94] In 1990 president Bongo had his longtime critic Joseph Rendjambé murdered in his hotel room in Port Gentil.[95] This kind of ungentlemanly behavior reveals president Bongo's centaurian side, which has helped him to keep his opponents, if not completely silent, then at least cautious of their words. An Amnesty International report on Gabon under the Bongo regime charged his national police forces with "torture by water immersion, electrical shock, and tobacco poisoning" as well as forcing prisoners to wear leg irons for long periods without medical attention, and other human rights violations.[96]

The "human" face of the Bongo regime is illustrated by his willingness to cooptate dissenters, when possible, into the ruling clan. For example, one of his more eloquent critics was Pierre Claver Maganga-Moussavou, whose *Does Aid Help?* (1983) suggested that the French were using the cooperation accords and economic assistance programs to further exploit Gabon—with the complicity of the Bongo regime. In 1990 president Bongo appointed Maganga-Moussavou as his economic adviser, and as a contributor to the writing of the Fifth Development Plan. Petroleum wealth provides president Bongo with this kind of option, and in the case of Maganga-Moussavou, illustrates Giacomo Luciani's assertion that "the solution of manoeuvering for personal advantage within the existing set-up is always superior to seeking an alliance with others in similar conditions."[97]

POLITICAL CHANGE IN THE APRES-PÉTROLE

Gabon is not the only dictatorship in Africa. But what is a special if not unique feature of this dictatorship is the source of its revenues. A dictatorship that lives by oil will die by oil. The inverse relationship between economic rent and political democracy—which liberates the state from giving representation to a society that does not pay taxes—will also open the floodgates to opposition and revolution when the oil revenues run dry.

There is a parallel here between Mahdavy's observations on economic development in Iran an our own on political development in Gabon. Mahdavy, in his seminal work, observed that during the period 1950-1953, when Iran lost virtually all her oil revenues in the struggle for nationalization, the effect was "contrary to the expectations of many Western observers ... stimulating to the Iranian economy."[98] Because the prices and overall costs of previously imported commodities rose (as scarcity set in during the boycott) substitution of domestic products for marginal imports became profitable. "In textiles alone, in these three years new factories with 110,080 spindles and 1,600 weaving looms, were installed."[99]

This is the null-hypothesis for rentier theory: If a sudden increase in rent decreases economic development, then a sudden decrease in rent should increase it. Strange as this may sound—for the usual impression is that oil money is *per se* an economic windfall—the solution to many of the developmental problems associated with a rentier economy is that oil revenues must decrease rather than increase. Economic development in a rentier state requires cessation of rentierism, rather than increased allocation or reformed spending policies. Since one of the underlying assumptions of the present work is that the economic and the political are materially interdependent, the same, we suggest, may hold true for political democracy to ever arise in such a state.

The hypothesis that a decrease in Gabonese petroleum rents will increase the probability of democratization is warranted by the recent, post-1986 efforts by opposition groups within and without the country to end the Bongo regime. So far we have discussed the kinds of methods by which president Bongo has managed to consolidate his hold on the state. But in a dynamic world, it is equally important to understand the processes of change. No regime, after all, lasts forever.

In 1986, President Bongo accepted the terms of an IMF-sponsored austerity plan. Oil prices had been steadily falling for years, and with them, Gabonese government revenues. In order to continue his ambitious development projects, particularly the Transgabonese Railroad, president

Bongo had borrowed heavily from public and private international sources, landing his country in the pits of a severe debt crisis. The payments on the increasingly heavy national debt only added to the fiscal problems of a government budget already suffering tremendous deficits caused by falling oil revenues resulting from the oil-price crash of the 1980s. "Times are hard" president Bongo said in an election speech. "We are in a financial Ramadan."[100] The implicit meaning of this message was that he was going to put his nation on a financial fast.

In its first years, the IMF austerity plan cut $184 million from the government budget, froze wages for civil servants and the public sector at 1985 levels, and placed restrictive price controls on essential commodities while removing protective subsidies from others.[101] Transport and housing allowances for civil servants were cut. All of these cuts were to the kinds of programs that—while perhaps wasting money from an economic stand-point—were the meat-and-potatoes of president Bongo's political cooptation of dissent.

As the years of economic austerity dragged on, various opponents of the Bongo regime began to gain political momentum. For the first time since President Bongo had come to power, his people acquired an influence which they had never known as tax payers. It was as if the end of government allocation had to be somehow "justified." President Bongo, the man who presided over the single largest economic expansion ever in the history of Gabon, was now being held accountable for its most devastating economic decline. As the austerity plan cut government spending by a quarter, then a third, then a half, the Paris-based MORENA opposition party grew in influence. In 1986 president Bongo had invited two key MORENA members (André Nbaobamé, the party spokesman, and Parfait Anotho Edowisa, its secretary general) to return from exile in Paris. Both men were then elected to the central committee of the PDG and thereby silenced.

By 1989, however, the hardship caused by the IMF austerity plan was causing severe civil unrest. For example, in September several disgruntled Bapounou (Punu) and some government ministers and aids to the president attempted to overthrow the regime. Their coup, which involved an assassination attempt on the president, failed. Yet it was soon followed by another coup attempt in November. This second effort also failed. According to Bongo's Minister of Information, the *Garde Présidentielle* and other security forces in the country had kept the coup-plotters under close observation for some time before having them arrested. The coup attempts demonstrate something about the political instability of this ethnically divided regime. When the petroleum revenues which had soldered the national unity suddenly disappeared, and the allocation state no longer allocated as it once did, popular discontent and the chance for anti-regime activities increased. In 1989, President Bongo, responding to the threats, decided to enter into discussions with key MORENA opposition leaders, including Paul Mba Abessolé, who returned to Gabon in the fall of that year and inaugurated the process of political change and liberalization.

"Anyone who dares put disorder in this country," president Bongo warned Abessolé's followers in a speech commemorating the twenty-ninth anniversary of Gabonese independence, "will find that I am in power, and that I will never tolerate any one of them."[102] Such strong words only revealed an underlying fragility in the power structure. In late January, teachers, students, and hospital workers went on strike, demanding better pay. They were soon after joined by other workers in a general strike. Citizens of Gabon took to the streets, attacking government buildlings and overturning cars and setting them aflame. The riots continued into February, as policemen deserted their lines to join the protesters. "The army is on strike. Everybody is on strike!" proclaimed one opposition spokesman, Adrien Nguemah Ondo.[103] One French official who was posted in Port-Gentil commented that "[t]he streets are empty but for isolated gangs of youths."[104]

Abessole's return to Gabon came at a critical moment in the history of Gabon, one where a conjunction of high volatility and low solvency forced the regime to bend to pressures for progressive change. President Bongo announced that he would hold a conference to discuss democratic reforms in the country, including the possibility of a return to multiparty elections, and the scheduling of a new presidential contest; all of which was provoked by the outbreak of civil unrest in Port-Gentil, itself conditioned by the decrease in government oil rent allocations.

Oilworkers at the Gamba, Lucina, and Rabi-Kounga fields did not strike. And even though airport workers at Port-Gentil did walk out for 48 hours, their strike had little effect on the oil companies' operations as the companies were able to resort to sea transport of their staff and equipment. The petroleum derricks were designed for operations in an independent environment, with their own fleet of aircraft, power supplies, and telecommunications networks. So in a demonstration of their "enclave" features, life in the offshore oil derricks continued as usual, while onshore facilities shut down and social services came to a standstill.

"We need new political structures to adapt to a new era," President Bongo told a joint conference of the PDG and parliament in March, after the riots had stopped and order had returned to the country. The conference was held in a hall guarded by soldiers of the elite presidential guard. "Together we are going to build a stronger, fairer and more democratic Gabon on the new basis."[105] Bongo agreed to review salaries for 40,000 government employees, and called for the dismantling of the PDG and its replacement by a new, broader-based structure that he called the Gabonese Social Democratic Rally. Two days after his speech, however, bank and insurance company employees called a wildcat strike over pay; then university teachers demanded better conditions and boycotted classes. Junior doctors at a teaching hospital in Libreville walked out to claim payment of their December wages. Then a wildcat strike erupted at the country's only flour mill, which, according to *l'Union*, threatened to lead to a shortage of bread in the nation's capital.[106]

While the offshore oil derricks were islands of stability during this time of crisis, staffed by foreign workers and supplied by independent means, the onshore facilities became embroiled in the labor unrest. On March 21st, oilworkers at the country's only refinery threatened to shut down the SOGARA plant at Port-Gentil unless an across-the-board wage increase was granted by Elf-Gabon. Being government employees, the refinery workers had had their wages frozen since 1986 as part of the IMF austerity plan.[107] Their threatened shutdown provoked president Bongo to clamp down on Port-Gentil with a ten-hour curfew and an order banning any gathering of more than five people in the oiltown streets.[108] It is difficult to overestimate the influence wielded by the oilworkers, whose one-week strike staged in the early 1980s had cost Gabon an estimated one billion CFA francs per day. After a week, the government and Elf agreed this time to meet some of the workers' demands. The regime ordered a refund on a wage tax that had been levied in the previous year. President Bongo also cut the costs of health services and lowered water rates. He told strikers that an across-the-board wage increase would be negotiated for the following year and, in so doing, put an end to the refinery walkout.

But in May, riots erupted once again. This time the riots were in response to the assassination of opposition leader Joseph Rendjambé in a hotel room in Port-Gentil. In retaliation to what protesters saw as French complicity, a French embassay official, M.Duffau, was taken hostage by supporters of Rendjambé. Immediately president Bongo imposed another curfew, while his residence in Port-Gentil was vandalized by angry mobs.[109]

Elections had been scheduled for April of that year, in accordance with the reforms promised by president Bongo to open up Gabon to competitive multiparty elections. But when opposition leaders began to appear to have a potential advantage, apparently it was judged expedient to have Rendjambé murdered. The ensuing riots were the worst ever to hit Gabon. One thousand people took to the streets of Port-Gentil in defiance of the government curfew. They burned down the government-run hotel where Rendjambé's body had been found, and also burned down a nearby movie theater that president Bongo had built for the 1977 OAU conference. Then they set out on a rampage through the streets, looting the few remaining shops and burning buildings at random. André-Dieudonné Barre, the director of Shell Gabon, complained that "[t]he rioters systematically destroyed the infrastructure that we have invested many years to create."[110] The French consulate in Port-Gentil was also burned down, prompting the French to once again invoke their defense accord with Gabon and to call in 500 troops from their Foreign Legion. Libreville saw the stoppage of traffic as crowds shouting anti-regime slogans and demanding multiparty reforms massed around the presidential palace.

In a surprisingly familiar scenario, French troops were stationed at key positions throughout the capital city and especially around Port-Gentil. They arranged a convoy of buses from the giant Elf facility to the airport, where military cargo jets were ready to airlift 1,800 expatriates who worked in the country. Elf Gabon evacuated all but 50 of its 600 French staff, leaving the bare minimum behind to continue token production from the paraffin-rich fields (which would be impossible to restart if totally shut down). General Paul Poncy, in charge of the French troops, said "[w]e'll avoid bloodshed and attempt to disarm the mobs when the moment comes."[111]

The events of May 1990 exposed the neocolonial mechanisms that interconnected post-1986 socialist France under François Mitterrand with the status-quo Gaullist institutions of Bongo's Gabon. For the first time in the country's history, the French oilmen of Gabon were forced to stop production. Angry mobs took ten Elf and Shell executives hostage and demanded an end to the Bongo regime.

Information Minister, Jean-Rémy Pendy-Boutiki told reporters that Gabon was losing 50 million dollars a day in export revenues: "If oil production is interrupted for more than 10 days, the whole country will be on its knees."[112] The Mitterrand administration found itself confronted with institutionalized arrangements dating back to the days of de Gaulle. The French state controlled some 65 percent of Elf through direct ownership and holdings by state-owned firms. Gabon accounted for nearly a quarter of Elf's worldwide oil production in 1989. What was at stake for Gabon was its entire economic well-being. What was at stake for France were company losses estimated by Elf officials at 2 million dollars a day. What was a stake for both was a long, profitable relationship between the Bongo regime and

France. Invoking the defense accords yet again, and calling out the French troops—reminiscent of 1964—preserved Franco-Gabonese relations as they existed. Eleven hostages (when including the consul) and the lives of 1,800 French citizens were at stake. But one has the sense that having established themselves so deeply in Gabon, the Frenchmen responsible for conducting Franco-Gabonese affairs would have used military force to protect their interests anyway. French troops evacuated 200 people from Rabi-Kounga field and were stationed in and around the oil facilities along the coast.

Eventually the 10 oil executives who had been taken hostage were released unharmed. This came after a detachment of French troops was sent in to deal with the rioters outside Shell headquarters, where (according to one of the hostages) Shell Gabon director Roland Toulouse had been held in administrative offices.[113] Meanwhile president Bongo went on French television that evening threatening to drop the services of Elf if it did not return its evacuated workers and resume production immediately. According to president Bongo, law and order was being adequately enforced by the military troops, and Elf's decision to airlift its employees out of Gabon had been, in his words, "completely unjustified."[114] Elf subsequently announced the next day that it believed "sufficient security measures were being taken in the Port-Gentil area to start production back up."[115] Company spokesmen indicated that the presence of French soldiers in Gabon was a factor in its decision to resume production, after several hundred more soldiers were sent to assist the already stationed attachment in Port-Gentil. Residents said that French tanks destroyed barricades erected by insurgents during the night, and fought with rebel "troops" (read: civilian protesters) throughout the night in the African district of Port-Gentil. Two opposition leaders—Pierre-Louis Agondjo Okawé and Augustin Mbounah—were reported mysteriously "missing" by late that evening. The remaining 500 French residents in Port-Gentil were grouped, under military protection, at a hotel in the Elf compound, while French troops effectively crushed all anti-government resistance in the country's oil capital by Thursday night.[116]

Details of the clamp-down included some tales of atrocity, as for example the reports by Radio France Internationale that Bongo's presidential guard had fired indiscriminately on crowds of protesters. These unarmed victims had obeyed the government's call to lay down their weapons an to demonstrate peacefully. French troops protecting French citizens withdrew to the sea, "leaving the presidential guard to do its work concerning the population."[117] Prime Minister Casimir Oya-Mba denied these charges, claiming that the GP had been sent to Port-Gentil to "maintain order" and not to "massacre" civilians.[118] The army reported that one civilian was killed and six wounded in the worst outbreak of violence to hit the country. Other reports said that six people were killed and eleven wounded. One government soldier was reported to have been beaten to death during clashes that day with demonstrators in Port-Gentil. Other reports claimed that a young girl was shot and killed in the violence.[119] President Bongo announced over a nationwide broadcast that he was ready to re-establish order "by all measures necessary."[120] He imposed a state of siege in the Ogooué-Maritime province, which includes Port-Gentil, and later sent word via General Paul Poncy that this state of siege would remain in effect for 12 days. Poncy told reporters that more than 40 people, including three opposition leaders had been arrested.[121] France meanwhile sent 500 troops to Port-Gentil to reinforce its position and protect the oil facilities, which restarted production that week.

After the riots were quelled, president Bongo ordered an investigation into the death of Joseph Rendjambé. His followers claimed that he was murdered by lethal drug injections after entering a hotel room with a woman from the Ivory Coast who had been hired by president Bongo. But the final report issued by the Bongo administration maintained that his death had occured through "natural causes" and that "the syringe marks on his stomach were caused by the injection of drugs to treat diabetes."[122]

Elections were rescheduled for September of 1990, with each of the newly legalized opposition parties receiving— in accordance with the regime's normal mollifying tactics—an allocation of 20 million CFA francs (approx. $50,000). Those parties that won a seat on the newly constituted National Assembly additionally were allocated 30 million CFA francs and a jeep.[123] The government of Gabon was charged with ballot-stuffing in these multiparty elections, as well as widespread fraud and intimidation of voters. Election results in the most hotly contested voting districts were inexplicably delayed, and in 32 districts were subsequently annulled. Not surprisingly, the president's PDG was the winner of these multiparty elections, holding on to 44 of the 58 contested seats. By-elections held in October for the annulled results produced an additional 18 seats for the PDG, which gave it a parliamentary majority (62 out of 120) in the newly constituted National Assembly. Meanwhile the case allocations to the new opposition parties guaranteed a working majority on all matters governmental. "The PDG received pledges of support that should allow it to control the legislature."[124]

The political unrest precipitated by the fall in oil prices and the IMF austerity plan was briefly palliated by the elections and by a stroke of good luck for the Bongo regime. In 1990 the price of oil increased in response to the Persian Gulf crisis, which made it possible for president Bongo to double his national budget for fiscal year 1991 to around two billion dollars ($1.96 billion, or 490 billion CFA francs). The oil revenue windfall from the Gulf crisis generated about half the total budget for that year.[125] Also the 6 seats won by Rendjambé's PGP and the 15 seats won by Abessolé's MORENA channelled much of the most dangerous criticism of president Bongo through the muffling mechanisms of parliament.

But when oil prices fell, the government was forced to cut its crude oil exports and, subsequently, its prime source of external rent. On February 18th, Gabon announced that it would cut its production by 2.5 percent in order to meet an OPEC quota.[126] On Tuesday of the following week a general strike was called by opposition parties, paralysing Port-Gentil in a night of sporadic violence. The Elf-Gabon oilworkers went on strike first, drawing four government ministers in from Libreville to negotiate with them. Elf spokesman Robert Chabert reported that some of the company's producing facilities had been sabotaged. Offshore crude oil production was stopped by "the personnel of the rigs" who posted a 47-point list of demands—including one for their promised pay increases. Meanwhile a coalition of opposition parties called for a general strike until the government allowed the "safe return of political exiles" and adopted fair procedures for a presidential election.[127]

This strike lasted only a week, as Elf threatened to permanently dismiss all workers who stayed on strike for breach of contract. The general strike lasted only one day, as business in Port-Gentil reopened and life in the oil city returned quickly to normal. The government never did agree to return the exiled leaders, nor to any other demands regarding the presidential elections—except for the fact that there would be an election scheduled for the office of the presidency, details pending. The presidential election was eventually scheduled for the winter of 1993, and president Bongo was generally expected to win it. After all, didn't he always? *What the events of the past six years had demonstrated was that while declining petroleum revenues have threatened to overthrow the Bongo regime, a quarter century of rentier economics and depoliticization has not empowered any social group (party, union, etc.) to successfully replace him.* And cooptation of the oilworkers only demonstrates how the solution of manoeuvring for personal advantage within the existing setup is always superior, in a rentier state, to seeking an alliance with others in similar conditions. For the time being, Gabon appears to be at a standstill, with the forces of change battering against those of invariance.

At such times the common solution has been to vent aggression on foreign workers. Rising xenophobia in Gabon resulted in massive expulsions of foreign workers in the fall of 1992. State paramilitary gendarmes were reported to have scoured poor areas of Libreville rounding up Nigerians and Camerounians who were not carrying proper papers. "Those arrested suffered humiliation and physical brutality" charged opposition leaders in PGP. François Ondo Nzé, the chairman of the Gabonese Human Rights League (LDGH) claimed that "tens of thousands of

Nigerians with valid resident permits had been stripped of their belongings."[128] Like a pressure valve being released, the maltreatment of outsiders can only temporarily relieve the explosive forces pent up in a society repressed under one-party personal rule for more than twenty-five years.

On December 5th, 1993, President Bongo faced a dozen challengers in his country's first even presidential election for the office of the presidency. The election was to involve somewhere on the order of 450,000 eligible Gabonese voters, who had an eleven-hour period (from 7 am to 6 pm) in which to go to one of the hundreds of polls in the country, "although the closing time could be delayed if there was violence" reported AFP in a telling statement of the government's uncertainty.[129] Lines were visible at the 230 polling stations in Libreville, which alone holds around a third of the eligible electorate. Foreign observers from the Organization of African Unity (OAU), the International Commission of Jurists (ICJ), the African-American Institute (AAI), and the African National Congress (ANC) issued a joint statement on December 6th to the effect that "they had found widespread flaws but no evidence of deliberate fraud."[130] By December 7th, French radio reported that provisional interim results released by the Election Commission in Libreville showed Omar Bongo leading in the polls. There were organizational problems at the ballot boxes, and results which had been promised by December 6th were officially delayed by a week. Meanwhile the opposition formed a coalition, the Convention of Forces for Change (CFC), which warned that there would be massive unrest if Bongo was proclaimed the outright winner. The rules stated that if no candidate among the 13 won an outright 50 percent of the turnout in the first round of voting, then a second run-off election would be held a fortnight later on December 19th. CFC official Jules Aristide Bourdes-Ogouliguendé told a tense news conference "[w]e will destroy, we will seriously lay waste!"[131]

On December 10th, the National Electoral Commission declared president Bongo the winner of Sunday's race on the count of the first ballot, with 51.7 percent of the vote. This was surprising in the sense that, with so many candidates, a second ballot seemed statistically assured. But it was not surprising in the sense that, yet again, Omar Bongo won the presidential election and retained personal power in Gabon.

Paul Mba Abessolé, the leader of the opposition parties, pushed ahead with forming his own government on December 11th, hours after he declared himself the winner of the disputed election.[132] The following day four major unions ordered a general strike after hearing reports that the government had killed three people. The opposition members of the National Assembly spent the day assembling what they termed a "fighting government" to wrest control of Gabon from its autocratic ruler. The unions, for their part, said that their strike would continue until the government conceded to "the democratic voice" of the Gabonese people.[133]

Although it is impossible to forecast the future with much precision, there are good reasons to believe that President Bongo will not remain in power. One of these is that the men who "made" him are themselves disappearing from the scene. For example, Jacques Foccart, who had been so instrumental in the shaping of Gabonese political actualities, found himself expunged from office during the Mitterrand presidency. President Chirac had brought Foccart back as an advisor, but no one seriously imagines that the *homme de l'ombre* will return to his former power in African affairs. There is also a sense that the long-term relations President Bongo has formed with the French right are beginning to fail him, and it may not be long before he becomes a political liability rather than an economic asset.

Already the French government has substantially privatized its state oil company, Elf, retaining a mere 15 percent for itself. As it divests itself of Elf and, in the process, the subsidiary operations of Elf-Gabon, who knows what kind of military commitment France will be willing to make on behalf of an autocrat who will not step down from power?

Another reason that Omar Bongo may not remain in power is the fact that the economy of Gabon is becoming less "rentier" with each passing year. As the rentier state is a subset of a rentier economy, the decline in the latter forecasts the fall of the former. *Unfortunately the fall of*

Omar Bongo may only precipitate the rise of someone much worse. As deforestation and depletion of nonrenewable natural resources drain Gabon of her precious dowry, the state must prepare itself for ecodisaster and insolvency. If things were bad for Gabon during its heydey of rentier riches, wait until it is just another poor African country.

Ultimately, however, nothing can be truly free until it is torn from those things which bind it down in perpetual obligation or dependence. The Irish bard W.B. Yeats wrote "[n]othing can be whole nor sole until it is rent." He may have been referring to the relationship between England and Ireland, but the same could be said for France and Gabon. In order to become what it purports to be, somehow the Republic of Gabon must escape from the dominance-dependence relationship in which it finds itself with its former colonial metropole.

NOTES

1. Carl Jackson and Carl Rosberg, *Personal Rule in Black Africa* (Berkeley: University of California Press, 1982), p. 144.
2. Ibid., p. 143.
3. Ibid., p. 156.
4. Ibid., p. 143.
5. Pierre Péan, *Affaires Africaines* (Paris: Fayard, 1983), p. 20.
6. Joan Edelman Spero, "Dominance-Dependence Relationship: The Case of France and Gabon" (Ph.D. dissertation, Columbia University, 1973), p. 1.
7. Péan, *Affaires Africaines*, p. 129.
8. Max Weber, *The Theory of Social & Economic Organization*, 1947.
9. Karl David Patterson, *The Northern Gabon Coast to 1875* (Oxford: Clarendon Press, 1975), p. 17.
10. Ibid., p.68.
11. Nicolas Métengué N'Nah, *Economies et Societés au Gabon dans la Première Moitié du XIXe Siècle* (Paris: Harmattan, 1979), p. 15.
12. Patterson, *Northern Gabon Coast*, p. 43; see also: David Gardinier, *Historical Dictionary of Gabon* (London: Scarecrow Press, 1981), p. 178.
13. Patterson, *Northern Gabon Coast*, p.34.

14. For a detailed account see Thomas Pakenham, *The Scramble for Africa: The White Man's Conquest of the Dark Continent from 1876 to 1912* (New York: Random House, 1991).
15. Albert Schweitzer, "On the Tracks of Trader Horn," *African Notebooks* (Indianapolis: Indiana University Press, 1939).
16. Pakenham, *Scramble for Africa*, p. 632.
17. Edmund D. Morel, *The British Case in French Congo: The Story of Great Injustice, Its Causes and Its Lessons* (London: William Heinemann, 1903), p. 185.
18. Catherine Coquery-Vidrovitch, *Africa: Endurance and Change South of the Sahara* (Berkeley: University of California Press, 1988), p.37.
19. Ibid., p. 37; cf., G. Sautter, *De l'Atlantique au Fleuve Congo: Une Géographie de Sous-Peuplement*, (Paris: Mouton, 1966).
20. Virginia Thompson and Richard Adloff, *The Emerging States of French Equatorial Africa* (Stanford: Stanford University Press, 1960), p. 88.
21. Aimé Césaire, *Monthly Review Press* (1972), originally published as "Discours sur le colonialisme" by *Présence Africaine* (1950); cf., *African Aims and Attitudes: Selected Documents* (Cambridge: Cambridge University Press, 1979), p. 51.
22. Thompson & Adloff, *Emerging States*, p. 26.
23. Naval Intelligence Division, *French Equatorial Africa* (London: N.I.D., 1942), p. 281.
24. Coquery-Vidrovitch, *Endurance and Change*, p. 95.
25. Thompson & Adloff, *Emerging States*, p. 59.
26. *Grand Council Debates*, November 24th, 1947; cf., Thompson and Adloff, *Emerging States*, p. 26.
27. Terry Francis McNamara, *France in Black Africa* (Washington: National Defense University, 1989), p. 34.
28. Michael C. Reed, "Gabon: a Neo-Colonial Enclave of Enduring French Interest," *The Journal of Modern African Studies* 25, 2 (1987):288.
29. Ibid., 289.
30. Brian Weinstein, *Gabon: Nation Building on the Ogooué* (Cambridge: MIT Press, 1966), pp. 165-166.
31. James Barnes, *Gabon: Beyond the Colonial Legacy* (Boulder: Westview Press, 1992), p. 30.
32. Text from de Gaulle's opening speech; cf., McNamara, *France in Black Africa*, p. 49.

33. J.H. Aubame, "La Conférence de Brazzaville" in *Afrique Equatoriale Française* (Paris: Encyclopédie Coloniale et Maritime, 1950), p. 186.

34. Spero, *Dominance-Dependence*, p. 295.

35. Ibid., p. 197.

36. Ibid.

37. Barnes, *Gabon*, p. 33.

38. Gardinier, *Historical Dictionary*, p. 133.

39. *Bwiti* was a masculine secret society reinterpreted by the Fang to become a syncretic religion mixing their tribal rituals with Roman Catholicism. The charges of cannibalism may have been simply a tool used to discredit the unruly M'Ba. However, Brian Weinstein also explains that the term "to eat" had a symbolic meaning far more common than its literal one.

40. Weinstein, *Nation Building*, p. 172.

41. Charles and Alice Darlington, *African Betrayal* (New York: David McKay Co., 1968), p. 7.

42. Spero, *Dominance-Dependence*, p.297.

43. John Ballard, "The Development of Political Parties in French Equatorial Africa" (Ph.D. dissertation, Fletcher School, 1964), pp. 289-349.

44. Gardinier, *Historical Dictionary*, p. 93.

45. Barnes, *Gabon*, p. 35.

46. Ibid., p. 39.

47. McNamara, *France in Black Africa*, p. 190.

48. Pierre Péan, *L'homme de l'ombre: Eléments d'enquête autour de Jacques Foccart, l'homme plus mystérieux et le plus puissant de la Ve République* (Paris: Fayard, 1990), p. 125.

49. Ibid., p. 152.

50. Ibid., p. 213.

51. Ibid., p. 199.

52. Ibid., p. 207.

53. Ibid., pp. 213-234, "Activités occultes".

54. Quoted from General Massu; cf., Pean, *L'homme de l'ombre*, p. 224.

55. SAFIEX was created by Foccart on October 16th, 1945, while France was under German occupation, and "remains to this day the base of private professional activity for Jacques Foccart." (Péan, *L'homme de l'ombre*, p. 132).

56. McNamara, *France in Black Africa*, p. 188.

57. Remember that these had been colonies and therefore

did not require the same diplomacy which independent states would have required.

58. Spero, *Dominance-Dependence*, p. 323.
59. McNamara, *France in Black Africa*, p. 192; *cf.*, Brigitte Nouaille-Degorce, "Bilan politique de la cooperation," *Projet*, May (1962), pp. 551-552.
60. This account, given in Charles Darlington's memoirs of his experiences as U.S. ambassador to Gabon during the time of the coup, has also been supported by Pierre Péan.
61. Péan, *L'homme de l'ombre*, pp. 306-307.
62. Darlington, *African Betrayal*, p. 166.
63. Péan, *L'homme de l'ombre*, p. 310.
64. Darlington, *African Betrayal*, p.153.
65. Péan, *Affaires Africaines*, p. 50.
66. Omar Bongo, *El Hadj Omar Bongo: Par Lui-Même* (Libreville: Editions Multipress, 1983), p. 9.
67. Ibid.
68. Péan, *Affaires Africaines*, p. 46.
69. Ibid.
70. Maurice Delauney, *De la casquette a là jaquette* (Paris: Pensée Universelle, 1982).
71. Reed, *Neo-colonial Enclave*, p. 283.
72. Ibid., p. 64.
73. Ibid., p. 64.
74. Omar Bongo, *Gouverner le Gabon* (1968); *cf., El Hadj Omar Bongo: Par Lui-même* (Libreville: Editions Multipress, 1983), p. 15.
75. Ibid., p. 16.
76. Omar Bongo, *Dialogue et Participation* (Monaco: Paul Bory, 1973), p. 40.
77. Aicardi de Saint-Paul, *Gabon*, p. 27.
78. Barnes, *Gabon*, p. 49.
79. Ibid., p. 50.
80. Aicardi de Saint-Paul, *Gabon*, p. 36.
81. Barnes, *Gabon*, p.50.
82. Ibid.
83. International Labour Office, *Book of Labour Statistics* (Geneva: I.L.O., 1970), p. 310.
84. Ministry of the Economy and Finances, *Code Général des Impôts Directs et Indirects* (Libreville: Ministère de l'Economie et des Finances), Article 39.
85. Péan, *Affaires Africaines*, p. 146.
86. Ibid., footnote, p. 146.

87. *Code Général des Impôts*, p. 41.
88. Péan, *Affaires Africaines*, p. 154.
89. United Nations, *U.N. Statistical Yearbook* (1987); cf., Barnes, *Gabon*, p. 122-123.
90. Barnes, *Gabon*, pp. 52-53.
91. Péan, *Affaires Africaines*, p. 35.
92. Ibid., p. 36.
93. Niccolo Machiavelli, *The Prince*, chap. 18.
94. Péan, *Affaires Africaines*, chap. 1.
95. Barnes, *Gabon*, p. 65.
96. Amnesty International, *Gabon: Deni de justice au cours d'un procès* (1984); cf., Barnes, *Gabon*, p. 62.
97. Beblawi and Luciani, *The Rentier State*, p. 74.
98. Hossein Mahdavy, "Patterns and Problems of Economic Development in Rentier States: The Case of Iran," in: M.A. Cook, ed., *Studies in the Economic History of the Middle East* (Oxford: Oxford University Press, 1970) p. 442.
99. Ibid.
100. Peter Blackburn, "Leader of Tiny Oil-Rich Gabon Puts His Nation on Financial Fast " *Christian Science Monitor*, December 4th, 1986, p. 26.
101. *Xinhua General Overseas News Service*, August 17th, 1990.
102. *Xinhua General Overseas News Service*, August 17th, 1980.
103. "Gabon Oil Sector Spared in Recent Unrest," *Reuters* (March 2, 1990).
104. "Gabon Cut Off From the Rest of the World, French Sources Say," *Reuters* (February 26, 1990).
105. Jean-Loup Fiavet, "Gabon Puts Off Elections, Political Reforms Promised," *Reuters* (March 2, 1990).
106. "Gabon Faces New Wave of Strikes over Austerity Policies" *Reuters* (March 4, 1990).
107. Jean-Loup Fiavet, "Elf-Gabon Makes Offer in Pay Dispute with Oil Workers," *Reuters* (March 6, 1990).
108. "Gabon Clamps Curfew on Oil Port After Riots" *Reuters* (March 25, 1990).
109. "French Consul Taken Hostage in Gabon" *Reuters* (May 23, 1990).
110. *Times Newspapers Ltd.*, (May 28, 1990).
111. Gilles Trequesser, "Gabon, Rocked by Civil Unrest, Fears for Economy" *Reuters* (May 28, 1990).
112. Ibid.
113. "Elf, Shell Employees Freed in Gabon," *Platt's Oilgram*

News (May 29, 1990).

114. "Elf Gabon Restarting Production," *Platt's Oilgram News* (May 30, 1990),p.1
115. Ibid.
116. "Troops Crush Oil City Rebellion in Gabon," *Reuters* (May 30, 1990).
117. "Gabon clashes between Presidential Guard and government troops in Port Gentil" *British Broadcasting Corporation*, May 31, 1990 (ME/0778/ii).
118. Ibid.
119. Gilles Trequesser, "Dead Opposition Leader Becomes Symbol of Gabon's Long Crisis," *Reuters* (May 31, 1990).
120. *British Broadcasting Corporation, op. cit.*
121. Adeline Bommart, "Troops Crush Oil City Rebellion in Gabon," *Reuters* (May 30, 1990).
122. "Bits & Pieces" *Africa News* (June 11, 1990).
123. Francis Kpatindé, "Le Gabon entre Bongo et les Bûcherons," *Jeune Afrique* 1558 (November 7-13, 1990):28-29.
124. Barnes, *Gabon*, p. 67.
125. "Oil Revenues Boost Gabon's Budget," *Reuters* (December 15, 1990).
126. "Government to Cut Crude Oil Production by 2.5%," *British Broadcasting Corporation* (February 18, 1992).
127. Pierre Briand, "General Strike and Clashes in Port Gentil" *Agence France Presse* (February 25, 1992).
128. "Gabon Expells More Than 10,000 Nigerians," *Reuters* (October 21, 1992).
129. "Gabon Goes to First Pluralist Polls," *Arab News* (December 6, 1993):24.
130. "Foreign Observers Endorse Gabon Vote, With a Caveat," *New York Times* (December 7, 1993):A8.
131. "Gabon's President Heads for Victory," *Arab News* (December 8, 1993):20.
132. "Abessolé Goes Ahead with Govt Formation" *Arab News* (December 12, 1993):32.
133. "Troops Gun Down Three in Gabon," *Arab News* (December 13, 1993):1.

Chapter 4

The Dutch Disease and the Rural Sector

INTRODUCTION

An "allocation state" is a natural outgrowth of a rentier economy, one that is free to spend considerable amounts of money without having to resort to domestic taxation policy. The government under Omar Bongo has been able to allocate hundreds of billions of CFA francs to its various plans for social and economic development. But now that the oil boom is over, now that the Gabonese must survive on their development projects (rather than the other way around), there is some question as to whether or not allocations for developmental projects might have been confused with development itself? There is also some question as to whether or not rent allocations have hurt rather than helped agricultural development in this tiny forested country of less than one million inhabitants. This chapter will detail the evolution of state plans for rural development, plans which, as we shall see, are more reflective of the rise of the allocation state in Gabon than they are of a green revolution. It will examine how oil rentierism has contributed to what economists have called the "Dutch disease," or economic distortions resulting the decline of non-oil sectors, and in particular, agricultural decline.

ALLOCATION AS/AND RURAL DEVELOPMENT

The basic model for rural development in Gabon traces its roots back to the French Ministry of Cooperation's *Economie et Plan de Développement*, which was the prototype document for all future development plans. In this document the Ministry conceptually divided the economy of Gabon into a "modern" and a "traditional" sector. The former was principally a European-controlled sector that furnished around 90% of exploited forestry products, the totality of petroleum, almost all of the minerals, and whatever factories and other large commercial enterprises existed in the town during the early years of political independence. The latter was, in the words of the Ministry, "exclusively African ... one of cultivating activities (agriculture, light breeding, hunting, river and lake fishing), of artisanship, and small traders."[1]

Rural development planners recognized that a disequilibrium existed between the so-called "modern" townspeople and their more "traditional" countrymen. Rural development plans therefore were designed as an effort to correct this imbalance. In the Ministry's discourse of nascent postcolonial development, the "modern" sector was identified by the planners as being comprised of 49,700 Gabonese and 2,600 Europeans who, when combined, produced around 75% of the total value added in the economy (p. 14). Meanwhile, the "traditional" sector contained 208,000 non-salaried individuals who, by Ministry estimates, were responsible for generating a mere 25% of value added (p. 15). In other words, 80% of the economically active population—the "traditional" sector"—lived a life of relative deprivation.

Rural development can be understood as a concerted effort to improve the living conditions of the majority of Gabon's population. It was not as concerned with modernization *per se*. The French planners were principally concerned with the "very serious negative repercussions ... characterized by excessive urbanization" (p. 34). At the time of the printing of the third edition of the Ministry of Cooperation's prototype triennal plan, the port-cities of Libreville and Port-Gentil had drawn large populations of

30,000 and 21,000 inhabitants, respectively (p. 6). How to accomodate for this rural exodus in the absence of adequate urban employment and housing was the problem.

French planners set the tone for future development plans by expressing an almost fatalist complacency toward the rural populations of Gabon. "The rarity and the feeble qualifications of the work force doesn't permit for a rapid industrial growth," the French planners inform us (p. 6). Rather than directing effort at the modernization and/or industrialization of the rural populations, which was understood to be practically impossible, the French called for "integration for the rural world in development." Four objectives were established by the Ministry to accomplish this integration: (1) the organization of rural cultivation in accordance to the needs of the urban areas; (2) the development of infrastructure sufficient for the collection, conditioning, and commercialization of rural products; (3) the technical supervision of rural producers; and (4) the experimentation in culture crop diversification (p. 39). These efforts were aimed at the preservation of the "traditional" sector rather than its "modernization."

When you look at the actual amount of investment committed to agriculture by the Fonds d'Aide et de Coopération (FAC)—not including the money spent on roads which were built more for forestry companies than for small farmers—one soon discovers that despite all of the concern that was expressed by the Ministry about the disequilibrium of the Gabonese economy, the financial assistance for agricultural *développement* (58.5 million CFA francs), *animation* (32.5 million CFA francs), and *expérimentation* (17.5 million CFA francs) amounted to only around 12% of the 1962 program (p. 47). The 1963 program dedicated only 13% (p. 48).

Of course one could argue that the plan was only supposed to be an intermediate plan. But according to the French planners themselves the reason that a triennal program "qualifies as intermediary [is] precisely because it is destined to assure the transition attending the elaboration of the definitive five-year plan" (p. 38). In other words, the prototype three-year plan was expected to establish a precedent for the future five-year development plans.

One could also argue that the FAC assistance provided by the Ministry of Cooperation could be expected to provide proportionally less to agricultural production than to other aspects of the economy that produced *real* returns on investment (i.e., forestry, manganese, oil) for the French. Meanwhile one would expect to see the state budget providing proportionally more than its French counterpart to agricultural projects of strictly Gabonese concern. But this was not the case. Direct expenditure recorded in the state budget on the rural collectives (46.2 million CFA francs) and on the Caisse de Stabilisation des Prix du Cacao (3 million CFA francs)—two government instruments of rural development—represented less than 1% of development expenditures (p. 44).

When the *Premier Plan de Développement Economique et Social* was released in 1966, president Léon M'Ba prefaced the government's first official five-year plan by calling it a *"réducteur d'incertitude"* and dissuaded his people from expecting too much. In president M'Ba's words: "The plan indicates the route to follow, it does not do away with the need to furnish effort, but renders it more effective through better coordination." Nevertheless the Plan was turned into law (*Loi-Programme no. 1-66*, July 26th, 1966) and continued to remain in operation even after the death of Léon M'Ba and his replacement, Albert Bernard Bongo. So it is not entirely inappropriate to treat the general orientations and objectives of the Plan as official government policy, if only for the years 1966-1970.

Ultimately the First Development Plan admitted that rural development was *not* going to be the government's number one priority. The planners explained that while "in other African countries, agriculture forms the base of the economy, and it is upon its vigour that the general progress awaits," the strength of the forestry, mining, and petroleum sectors were supposed to "carry sufficient resources to permit the development of non-agricultural sectors" without, one is left to surmise, investing too much in rural development.[2] The Plan created several administrative structures to coordinate rural development: the Ministry of Agriculture; the Office of Agricultural Services; the National Office for Agricultural Commercialization; the

Bureau for the Development of Agricultural Production; and the Institute of Tropical Agronomical Research. Coordination between these different agencies was assured by a Committee of Rural Modernization. The Office of Agricultural Services was responsible for the "conception, control and execution of agricultural policy" through a network of subordinate services and bureaus (p.163). For example the Agricultural Production Service was responsible for executing development and popularization (*vulgarisation*) of development programs from an administrative standpoint, while the Bureau of Studies provided the technicians in charge of research and implementation. Each service and bureau was responsible for a particular region. For example, the liason between the Ministry of Agriculture, the Office of Agricultural Services, and the various regional services and bureaus was assured by an Inspector of Economic Regions and an Inter-Regional Commission. The National Office for Agricultural Commercialization was essentially responsible for marketing and distributing agricultural goods to areas that were lacking in private commerce, and additionally became responsible for sustaining food prices (with the exception of cocoa prices, controlled by their own bureaucracy). The government created an Institute of Tropical Agronomical Research, mostly on paper, to achieve its objective of scientific experimentation in crop diversification (pp. 163-5). The point that must be made with respect to all of the above was that *these administrative and para-administrative offices consumed the vast majority of public monies earmarked for so-called "rural" development.* Since the administrative apparatus revolved around political life in Libreville, money purpotedly invested in "rural" development in fact went to "urban" development.

The actual situation, in all fairness, was that the government had not inherited a rich countryside with a highly developed farming base to start with. Remember that Gabon is a densely forested territory, not well suited for traditional forms of agriculture. At the time of the publication of the First Plan, the government estimated that out of a total land surface area of 28,700,000 hectares in Gabon, a mere 0.46%, or 125,000 hectares, was under cultivation

(p. 166). This cultivated microsector of the economy produced less than 16% of the gross national product and did not produce a wide variety of autoconsummable goods (p.157). The main crops grown for export were cocoa and coffee, with small amounts of palm oil, peanuts, and pepper cultivated around the country.

Cocoa—which represented by far the most significant agricultural product in terms of volume—was principally grown in the Woleu N'Tem region in the northern part of the country. In fact, according to the results of a survey conducted by the government for the Plan, the Woleu N'Tem region was responsible for around 95% of all cocoa production in Gabon (p. 166). Therefore only one region benefitted from this agricultural crop. But cocoa production had remained virtually stationary for the previous decade (averaging a little over two-thousand tons per annum). Cocoa prices, moreover, had declined severely: from 188 French francs per kilogram in 1959 to 98 French francs per kilo in 1962 (p. 167). So even the Fang who lived in this region were not getting rich on cocoa exports.

Coffee production had increased substantially during that same time period, rising from 13 tons in 1953 to 830 tons in 1963. But it still represented a very insubstantial cash crop for the country, and was itself, like cocoa, principally grown in the Woleu N'Tem. With 2,550 hectares of cultivated coffee the Woleu N'Tem region was the number one producer in Gabon (p. 168).

Palm oils were traditionally pressed and consumed by populations in the interior, but only two commercial plantations (SOGABOL & UNILEVER) had even experimented with cultivation and production for export—exporting 473 tons and 313 tons of oil respectively in 1963 (p. 168). Rice had been introduced in 1947 in the Tchibanga region, and its production had increased from 93 tons in 1947 to 1,080 tons by 1965 (p. 169). Peanuts were first commercialized for export in 1948 in the N'Gounié region, later spreading to Nyanga. Their production (calculated "in-shell") was 541 tons in 1957, peaking at 793 tons in 1959, only to suffer a steady decline to 1965 production of 440 tons (p. 169). Finally, the cultivation of pepper was the enterprise of but one plantation, run by two European planters, which peaked at 9,800 kilograms in 1956. These Europeans were preparing to abandon the project by the time of the First Plan (p. 169).

The thing to keep in mind with the above cash crops is that they were, for the most part, introduced after the Second World War under the aegis of Fonds d'Investissement et de Développement Economique et Social (FIDES). Over the course of the time period 1948-1957, FIDES supplied 944,400,000 French francs worth of rural assistance for agricultural operations. After 1959 these projects received their aid from FIDES' successor, the Ministry of Coopération's Fonds d'Aide et de Cooperation (FAC). FAC officially set aside as much as 551,500,000 French francs for agricultural operations started by FIDES (p. 165). Without FIDES and FAC monies, it is unlikely that Gabon would have had any export crops of significance by the time of the First Plan.

Most of the cultivated land was used not for growing export crops, but rather for foodcrops (*cultures vivrières*) such as manioc, bananas, yams, corn, potatoes, taro, rice, and diverse vegetables like spinach and squash, as well as tobacco. According to a study conducted by the government for the First Plan, out of the 125,000 hectares of cultivated land, around 95,000 (77%) were used for the production of foodcrops (p. 169). Estimates compiled during this agricultural inquest show manioc roots covering around 42,000 hectares, and supplying local peasants with around 165,000 tons of this starchy staple of the African diet. Banana trees covered 13,500 hectares of cultivated land (in addition to growing wild in the forest) and supplied around 75,000 tons of fruit. Yams covered around 6,000 hectares, yielding 27,000 tons of food. Peanuts that were grown for local use rather than for export covered an estimated 13,500 hectares and supplied around 10,000 tons of food. Taros were grown on 4,200 hectares and produced 17,000 tons. Diverse vegetables were also grown on 14,000 hectares (p. 170).

Animal husbandry of traditional European stocks was severely limited by the deadly rainforest topography and by endemic tropical diseases. Despite these hardships, Franco-Gabonese ranchers and pastoralists managed to raise 43,000 sheep, 47,000 goats, 5,100 pigs, and 278,000 fowl, as well as 3,486 head of cattle by the time of the First Plan (p. 172). Much more important to meeting the daily meat intake of the Gabonese peasantry was wild carnage: "The animal resources of Gabon are and will remain principally hunting and fishing" the planners cheerfully write (p. 170). But these activities tended to concentrate along the rivers more than along the coast, where imported meats were increasingly common. And at any rate, as we all know, careless hunting would soon lead to the extinction of many popular game animals.

If total foodstuff production in Gabon in 1966 were divided, hypothetically speaking, and if it were distributed equally to each and every single Gabonese citizen, then a personal diet consisting of 733 pounds of manioc, 333 pounds of bananas, 120 pounds of yams, 45 pounds of peanuts, 76 pounds of taro root, and an unspecified amount of green vegetables (augmented by whatever fish or game could be gleaned from the natural surroundings) would have been the annual per capita intake. Dividing the farm animals would further result in one head of cattle for every one hundred persons, one sheep and one goat for every ten persons, and one chicken for every two persons.

If we were however to factor in a virtually nonexistent internal road network (that prohibited any such distribution of product); and if we were to factor in regional concentration of crops and breeds (that would have reduced the supposed diversity of this hypothetical per-capita diet); and if we were to then factor in the absence of cash wages for purchasing of supplementary foods; and if we were to factor also the limited number of commercial enterprises outside of Libreville, Port-Gentil, and other towns in which goods could be purchased *were* cash wages readily available in the countryside; and if, finally, we were to include other factors such as the poor housing, the small number of hospitals, the low level of education, and the endemic diseases of the rainforest interior ... then we would have a different picture of living conditions for ordinary rural Gabonese.

These were the conditioning factors of the rural exodus, into which the booming manganese, uranium, and petroleum sectors could be factored to achieve a loose representation of the rural equation in Gabon at the time of the First Plan: "The expansion of the modern sector accelerates the rural exodus, and accentuates the demographic disequilibrium of the countryside" lament the planners, "rendering agriculture less and less permeable to modernization" (p. 176).

So it is here, at the beginning of planned development, that Gabonese agricultural policy is for the first time defined as a spending policy. *Rural development was defined as government allocation.*

The government planners write, in a section with the identifying subtitle *"définition de la politique agricole,"* that unlike other African countries where rural agriculture is called upon to play a leading role in economic growth, "the agricultural policy of Gabon should tend to augment the diffusion of revenues from the modern sector to the rural world" in order to permit the rapid transfer of development (p. 177). This policy statement is a reminiscent of Mahdavi's claim that rentier states do not have developmental policies *per se*, only spending policies.

Orchestrating this rural "welfare" policy (for that is really what we are talking about here) would be a large government administration. This administration would be concentrated in the capital city of Libreville. The government would call this part of its agricultural policy *"action extensive"* which was defined as "the pursuit of the activities controlled by the administrative services resulting in a supervisory network" (p. 179)

The second part of the government's policy of rural development was called *"action intensive"* and meant actions that were "strictly delimited and carefully chosen ... to test over the course of the first plan the psychological reactions of the encadred peasants to a new activity" (p. 179).

In plain language, these government spending programs were rural welfare projects. Like Lyndon Johnson's "Great Society" they were expected to completely change the peasants's way of life (*mode de vie*) and were designed to progressively cover, by expansion and extension, entire zones in which the agricultural vocation would be practiced. Like the giant dilapidated American welfare "projects" that one sees today rotting around New York City, these rural Gabonese "projects" would suck up vast sums of government money without any return on investment. Most would eventually be abandoned.

In the region of Tchibanga, where rice cultivation had been introduced in 1950, the government allocated funds for experimental *riziculture* to the village of Akok. The Akok rice project is a good example of the early optimism of the planners. With assistance from the Chinese, who provided technical training and expertise, fifteen families were granted one-to-five hectares of land for rice cultivation in a showpiece agricultural project. Money was also provided for a training center in Akok, where methods of rice growing and harvest techniques were to be exchanged. The Akok project was intended to experiment with rice cultures on deforested land (pp. 188-191).

In the region of Woleu N'Tem, where cocoa and coffee had been successfully grown since the war, the government allocated funds for three sub-regional projects— *"poles of attraction"*—which were designed to coordinate activities of small villages and plantations into larger and more rational productive unities. It is interesting to note that the concept of "poles of attraction" had been developed by radical marxist economists to describe how underdevelopment was occuring in Africa. "Poles of attraction" were accused of concentrating trade, wages, technology, and capital in various towns and cities, excluding the majority of the rural African populations. But in the Gabonese case the government planners took the radical theory and turned it on its head, attempting to create "poles of attraction" rather that redistribute the wealth that was, according to the original concept, supposed to be concentrated in them.

Of course, there was less to these "poles of attraction" than met the eye. The project at Meléné combined 24 villages with a total of 6,673 inhabitants into a zone of animation, and created a rural *bourg* that would supervise and coordinate the commercialization of the basic cash crops as well as the food crops within the area. The project at Bolossoville combined the efforts of 2,442 active male and female rural workers around an experiment in crop rotation that would supposedly augment their cocoa and coffee export revenues and lead the way toward crop diversification. Finally, the project at Mitzic encadred 3,206 inhabitants from 18 villages which had been growing coffee and cocoa, introducing commercial peanut farming to the zone. All these projects established government-run centers at the villages so-named, which were to serve as leading "poles of attraction" for further rural development (pp. 191-201). A brief visit to any of these three sites would raise doubts about whether they had really been so attractive to capital and technology.

In the region of Haut-Ogooué, where manganese and uranium mining had severely disarticulated the modern from the traditional populations (government estimates held that 81% of the population—the "traditionals"—shared a mere 16% of the monetary resources of the zone) the First Plan designated several rural projects in and around Koula Moutou. Administratively grouping together some 56 villages with a combined population of 5,513 inhabitants (of which 3,293 were active agriculturalists), the project attempted to create two-hectare family plantations which would be supervised from the rural *bourg* Koula Moutou. This "pole of attraction" would be a source of technical assistance, coordination, and commercialization of the coffee, cocoa, and pepper crops introduced to this far-eastern region of the Gabonese rainforest interior (pp. 205-210).

In the region of N'Gounié-Nyanga, where forestry had been the main economic activity outside of traditional rural foodcrop production and animal husbandry, the government created a "pole of attraction" at Lebamba, regrouping four villages in which the inhabitants were purportedly of the same ethnicity. Diffusion of rice cultivation in this southwestern canton was orchestrated through the propagation of three rural centers at Mondo, Kanga, and Mabanga. Through these spotlight projects the government wanted to experiment with reducing the women's share of agricultural labor and to increase the men's share. According to the First Plan the traditional responsibility of women in agriculture was one of the reasons that men were leaving the rural areas for work in the modern sector. At the centerpiece of the effort stood the Centre d'Appui Technique, which was to distribute goods and services for the diverse actions (pp. 205-210).

In the Ogooué-Ivindo region the government committed money for the establishment of a research outpost for the rehabilitation of agriculture in the area (pp. 215-217). Money was also allocated by the government for two centers in Libreville and Medoneu for the cultivation of garden vegetables—which, we are told, "Gabon imports practically 95%" (pp. 210-213). An agronomical research center at Lebamba (pp. 213-215) was also earmarked for allocation, as was a palm oil plantation at M'Vily (pp. 217-218).

Altogether the government planned to spend some 793 million CFA francs on agricultural projects in Gabon between 1966-1970, with the allocations divided as follows: 334.1 million CFA francs for the projects at Woleu-N'Tem; 87.7 million CFA francs for the projects at Haut-Ogooué; 88.2 million CFA francs for the projects at N'Gounié; 22 million CFA francs for the projects at Ogooué-Ivindo; 50.7 million CFA francs for the Libreville-Medouneu vegetable centers; 144 million CFA francs for the plantation at M'Vily; 14.5 million for the Lebamba agronomical research center. Additionally, the government earmarked 237.7 million CFA francs for the creation of an industrial fishing sector, and 99.2 million CFA francs for the establishment of several animal breeding projects, for a grand total of 1.129 million CFA francs expenditure on "rural development" (p. 219).

The source of public finance for this rural development was to be provided by Gabonese public funds (452.6 million CFA francs), foreign public funds (442 million CFA francs), and private capital (253.3 million CFA francs), of which the latter category was principally invested in commercial fishing (p. 222).

When the *Deuxième Plan de Développement Economique et Social* was released in 1971 (*Loi-Programme no. 5/71*, October 15th, 1971) president Albert Bernard Bongo prefaced his first programme with the optimistic and ambitious promise that it would "increase production by more than a third, and individual consumption by more than half."[3] If this language seems a dramatic departure from the fatalism of the French planners who preceded him, or from the cautious reserve of the M'Ba government, then it shared with them a certain conception of the subordinate role of agriculture. The number-one priority of the Second Plan was defined as the "pursuit of extraction of the natural resources" of Gabon, highlighting tropical woods, minerals, and petroleum. Development of the rural economy was relegated on the list to priority number two (p. x). During the years that were to be covered by the Second Plan (1970-1975), the sudden explosion of oil prices was to prove that this prioritization had been, with hindsight, justified. The return on investments in the oil sector would be astronomical, while various agricultural investments could not prevent the rural sector from falling into a steeper decline.

President Bongo shared with his predecessors the superficial concern with economic dualism in his country (i.e., modern vs traditional) but he said that he believed the best way to produce material well-being was through increased investment in the profitable mining sector. So the Second Plan opens with a review of the economy from 1964 to 1968. In this review we are informed that while agriculture had increased at an annual growth rate of 2.9%, the mining sector had almost tripled during the same time period. Based on an index of 100 (1964) we are shown that agriculture would only reach 112.1 in 1968, while mining would stand at 301.2 (p. 1). Furthermore agriculture was the only sector of the economy forcasted by the planners to suffer an actual decline in the gross formation of fixed capital (p. 7).

One gets the sense, upon reading this and previous plans, that *rural development was perceived as a necessary but somewhat unfortunate burden for the government to bear, rather than an investment, or a potential source of real growth in the economy.* With young men leaving the farms in increasing numbers, and with even more of them predicted to abandon their homes in search of employment on the Belinga-Owendo branch of the Transgabonais railroad, the government set as its "primary objective" in the rural development program "to maintain a dynamic fraction of the population in the agricultural milieu" (p. xi). The secondary objective was to increase agricultural production.

Once again, the Second Plan, like the First Plan, treated rural development as if it were rural "welfare."

Government planners called for continued allocation to pre-existing projects, and increased investment in what it now called the "rural poles of development" (p. 67). Agricultural policy was redefined into three categories which, like new casks for old wine, gave the appearance of innovative policymaking while actually continuing administrative-and-project-oriented allocation policies from the First Plan.

The first of these three categories was the *vulgarisation* (popularization) of agricultural techniques around the basic commodities sector. Under the First Plan, FAC money from the Ministry of Cooperation had been granted for the development of cocoa in Woleu-N'Tem, coffee in Haut-Ogooué, Ogooué-Lolo, and Ogooué-Ivindo, and peanuts in N'Gounié. Under the Second Plan the government was going to commit 147 million CFA francs for these projects, whose FAC aid would terminate in 1971 (p. 68).

The second category of rural development was the program for the creation of *Centres de Paysannat*, which the government hoped would restore the nation's declining rural life through infusion of public funds into large state-run agricultural projects. In this section of the rural development plan, young peasants were to be contracted by the Ministry of Agriculture for a period of 10 years to work on a variety of grand public-works projects involved with the growing of food crops and the breeding of livestock. It was kind of like Franklin Delano Roosevelt's Civilian Conservation Corps (CCC) or Bill Clinton's Americorps program: i.e., direct government employment and training. If a peasant left his post before the ten years contract expired, then he was to be replaced by another peasant, so that in this way permanent settlements of young men would exist in Gabon producing food for the nation. The *Centres de Paysannat* included the fruit and vegetable projects created under the First Plan in Libreville and Médouneu. They also included the rice cultivation projects and the cattle ranch at Nyanga. There were several new projects, such as the Okoumé Ranch, which were established jointly with research centers (pp. 68-69). What is particularly interesting and relevant to the discussion at hand is that these centers were more than mere subsidization of a rural way of life. They represented an attempt to re-invent it through allocation. The *Centres de Paysannat* were not "traditional" peasant farming villages but rather "modern" agrocapitalist plantations strategically funded to curb the rural exodus and provide self-sufficient national food production. In some cases (e.g., cattle ranching) the effort represented a clear break with the past, and reflected a change in diet (e.g., beefeating) on the part of urban populations as much as it reflected a new *mode de vie* for the rural peasantry.

The third category of rural development defined by the Second Plan was the experimental research into the diversification of crops through the introduction of new cultures. The two projects already extant by the time of the Second Plan were the rice cultivation project at the village of Akok and the garden vegetable project at Tchibanga (see above). In order to popularize the experimental crops under study at these two facilities the government planned to conduct two studies, one theoretical, the other applied, over a period of ten months in two of the *Centres de Paysannat* (p. 69).

Complementary to all three of the above-listed Ministry of Agriculture categories were ongoing efforts to collect and analyze statistical information in the Ministry of the Rural Economy. This Ministry was made responsible for the collection of production, commercialization, and price figures preconisant with the regulations of the UDEAC in which Gabon operated. The collection of agricultural statistics was expected to cost somewhere on the order of 108 million CFA francs, representing more than half the funds provided to the Ministry of Rural Economy in the Second Plan (p. 69).

In order to finance all of these efforts at rural development, the planners knew that their realization would require massive allocation of public funds to construction, gross materials, and furnishings. Recurrent charges on the national budget would also be required for material and personnel expenses of administration. That is to say, there would be a large administrative cost drawing from public monies. Total investments calculated by the planners forecast expenditures totaling 1,795 million CFA francs for the projects of both Ministries, with an additional 720 million CFA francs for personnel and material expenses met by the recurrent budget (p. 70). Paying the salaries and office expenses for the bureaucrats in Libreville was going to cost, in other words, around one-third of total expenditures for this supposedly "rural" program.

Skyrocketing oil prices, resulting from the first oil shock, enabled the Gabonese rentier class to increase its financial commitment to rural development in the *Troisième Plan de Développement Economique et Social* (1976-1980). "After two plans that directed their efforts at local efforts" the planners boasted "the Third Plan is the decisive one, in so far as it will launch agriculture and breeding on a national productive platform."[4] In no uncertain terms the Third Plan promised that rural development would be its number one priority.

Several conditions triggered this response. For one, the previously insignificant part played by agriculture in the gross domestic product was, despite all the efforts mentioned above, on the decline. The contriburion of agriculture to the GDP showed a constant downward slope: 16% (1964) to 8.5% (1967) to 6.2% (1971) to 2.8% (1974): (p. 5, p. 82). Unless something were done to check this descent, agricultural production in Gabon would virtually disappear.

Also, the government began to worry about its dangerous reliance on non-Gabonese food suppliers. "Foreign traders were believed to hold a practical monopoly on sales of staple foods," reported *West Africa* in 1975, when during a food crisis these foreign merchants refused to sell their wares at government-mandated prices.[5] This event triggered the purchase of large amounts of food by the government, and the use of armed forces to distribute this supply to Libreville and other urban centers. President Bongo ordered the explusion of foreign traders, mostly small merchants from Togo, Cameroun, and the Congo.

A moment's examination of this food crisis is required. What happened in 1975 was that high oil prices brought higher oil revenues. Higher oil revenues in turn brought on higher prices for non-booming tradeable goods such as food. Inflation was running at a rate of 11.4 percent over the course of the previous year, with manioc and banana prices alone being responsible for five to six percent of that inflation.[6] This rise in food prices left many of the urban poor without the means to pay for their basic dietary needs. In order to quell urban unrest, the government decided to officially freeze prices, which the foreign merchants were unwilling to meet, and which "caused many to forsake the urban markets and dispose of their produce elsewhere, leaving the towns short of food."[7]

Food imports had doubled between 1970 and 1975. This reflected, on the one hand, new consumer tastes on the part of those Gabonese who benefitted from the oil boom. (Replacing traditional Gabonese staples such as starchy manioc with new staples such as French breads.) But it also reflected the decline in local food production, despite all government efforts to the contrary, caused by a "constant drain of labour and capital away from subsistence and export agriculture."[8] The result was a critical dependence of imported food, something that the government desperately needed to rectify.

Since this was the period when oil rents were flooding into state coffers at unprecedented rates, *the government's approach to solving the problem was to increase spending.* A little more than eight billion CFA francs were committed over the five year period of the Third Plan, to "investments in the food crop agricultural production program."[9] More than 59 billion CFA francs were to be spent on investments in the "program of industrial cultures and export cultures" (p. 95). More than 18 billion CFA francs were going to be spent on ranching (p. 102).

There were many new projects earmarked for allocation under the program for *cultures vivrières*. For example, there were: a banana plantation at Ntoum; a fruit and vegetable project at Lambaréné; another one at Franceville; two rice projects north of Tchibanga at Moukalaba and Ndendé; a marsh-vegetable project at Libreville; another at Port-Gentil; and so on (pp. 85-88).

There were also many new projects in the *cultures*

d'export program selected for government funding. For example, there were: an expanded cocoa project at Woleu N'Tem (Oyem, Bitam, Minvoul); a coffee project in the Ogooué-Ivindo region; another in Haut-Ogooué (Okondja); and another in Ogooué-Lolo (Lastourville); a peanut project in the N'Gounie region; a palm oil project at the village of Moabi; a coconut plantation at Maymba and at N'Toum; a hévéa (rubber) plantation in Woleu N'Tem; a sugar project, and so on (pp. 91-95).

The *élevage* program saw a new group of general projects such as the infrastructure and equipment facilities at Makokou, Koulamoutou, Minvoul, Bitam, Lebamba, Mouila, Moabi, Makoko, Médouneu, Franceville. Then there was the enlargement of the Okouma Ranch and the creation of new ranches at Lekedi and Franceville , and so on (pp. 99-102).

It is not necessary to go into each of these projects with any great detail, *because in August of 1977 the government was forced to scrap most of them anyway.* For it was during the Third Plan that the government of Gabon, having put itself on "queer street" by the unfortunate combination of exorbitant spending and falling oil prices, was placed within the restrictive confines of an IMF austerity plan.

The new Minister of Finance, Jérôme Okinda, announced that "dramatic cuts in public spending" were going to be implemented in order to stabilize the country's financial situation. No new projects would be launched. Those projects planned but not yet under way would be scrapped. And those plans already started would be halted.[10] The country's financial situation had been severely weakened by a combination of falling revenues and rising expenditures, resulting in a balance-of-payments deficit of 44 billion CFA francs by 1977,[11] and a debt service of 45 percent of the annual budget by 1978.[12] Falling revenues were caused in part by the U.S. dollar's depreciation (this being the currency in which oil contracts were paid). Rising expenditures on the Transgabonais railroad, as well as lavish expenditures on the OAU conference hosted in Libreville in 1977, and the construction of a second presidential palace, had combined with these falling revenues to create a veritable fiscal crisis.

For the next four years (1977–1981) the country did not follow up on its earlier rural development plans, but rather eliminated most of its promised expenditures in order to meet the requirements of the IMF austerity plan. In the language of the IMF, Gabon "took the bitter pill." Debt relief was provided. And when the price of oil skyrocketed in 1979 during the second oil shock, Gabon was able to refocus its attention on rural development in a new, intermediate plan of economic and social development.

Employing the rhetoric of preceding plans, the *Plan Intérimaire de Développement Economique et Social* (1980-1982) promised to keep rural and agricultural development as the "priority of priorities."[13] However, in a critical report published by the Ministry of Agriculture in 1980, the Ministry complained that rural-agricultural development never had been the main priority. On the contrary, the report pointed out that *the first two development plans allocated to rural development projects only 2.7% and 1.7% of total funds, respectively*. It also pointed out that most of the projects/priorities of the Third Plan had "never seen the light of day." The report reserved special criticism for the agricultural projects already undertaken. According to the Ministry of Agriculture, the government's plans suffered from an "incoherent nature" that had resulted in a "series of projects developed in isolation from general policy plans." Most of these agricultural projects did not address the traditional diet of the Gabonese, but rather channelled public monies into foreign-run and foreign-owned agroindustrial export operations, or "luxury" production operations (such as the Port-Gentil green vegetable gardens). Meanwhile food imports had grown to 25 billion CFA francs worth of imports. And all of the cash crops had performed poorly. For example, the coffee fields were projected to achieve target production of 10,000 tons in 1978–79; but actual production was 234 exported tons.[14]

Even the Ministry of Planning had to admit that the rural and agricultural sectors had suffered "*une stagnation désespérante*."[15]

"As with other sectors" the Intermediate Plan observes, "modern agriculture ... has suffered the consequences of the economic fluctuation over the first few years of the preceding plan" (p.93). The Ministry admitted to the observable differences between investment objectives of the Third Plan and actual "realized" investments over the course of 1976–1980. The food crop program, for example, was supposed to receive 6.3 billion CFA francs (in 1975 francs), but actually received a total of 7 billion CFA francs (in 1980 francs). With inflation running at around 15 percent a year, the latter figure represents less in real value than the former figure, although it is higher in nominal value. The agroindustrial cultures were supposed to receive 23.3 billion(1975) CFA francs and only received 13.3 billion (1980) CFA francs. Ranching was supposed to receive 5.9 billion (1975) CFA francs but only got 3.9 billion (1980) CFA francs. General measures promised for rural agricultural development at 2.3 billion (1975) CFA francs ended up with a realized 500,000 (1980) CFA francs in allocation. All told, the objectives of the Third Plan for agriculture were 37.8 billion (1975) CFA francs. But "realized" expenditures were 24.7 billion (1980) CFA francs (p. 93).

Despite the fact that the government had not met its past commitments, the Intermediate Plan announced increased expenditures for all of its agroindustrial operations in the amount of 31,025,000,000 CFA francs over the course of 1980-1982 (p. 99). What are we to make of this brash optimism? Once more we see the same projects earmarked for public funding. The palm oil projects were targeted to receive 9,993 million CFA francs; the Mitzic hévéa project (4,379 million); the N'Dendé rice farm (3,915 million); the coconut farm (940 million); the poultry farms (3,826 million); integrated ranching at Nyanga (2,361 million) and Lekabi (1,380 million): (p. 99). Meanwhile public investments in the integrated zonal operations were targeted for a total of 18,555 million CFA francs, in which: The N'Toum banana plantation would receive 3,806 million: The fruit-and-vegetable project at Franceville would get 1,677 million: The existing cocoa operations at Woleu N'Tem would enjoy 3,959 million in new allocations ... and so on (p. 103).

The key factor to be considered when trying to evaluate why the government had originally expected to increase its allocations to rural and agricultural development over the 1980–1982 time period is that in 1979, as a result of the fall of the Shah of Iran and panic on the world oil markets, the price of oil soared to historical summits, thereby pumping new life into an otherwise invalid state budget.

Or to use the logic of rentier theory, *public expenditure on rural development was a consequence of the source of Gabonese public revenues rather than the success or failure of rural development projects themselves.*

When the government was deeply in debt and forced to go on an IMF austerity plan, for example, public funds promised for rural agricultural development projects were cut. When the government enjoyed an economic recovery due to extraordinarily high oil prices, however, it promised greater spending on rural development programs. The Intermediate Plan can be viewed, in this light, as an attempt by the government to pump funds into programs that had been drained—to make up in 1980–1982 what was lost in 1976–1980.

In his prefatory remarks to the *Cinquième Plan de Développement Economique et Social* (1984–1988) the new Minister of Planning, Pascal Nzé, suggested that the new plan ought to learn from the past mistakes of its predecessors. "The second and third plans were not executed up to their ends" he pointed out: "[T]he one because the first oil shock generated public receipts massively superior to its previsions" and the other "because the fall in the price of oil" combined with "uncontrolled anticipatory spending" forced the adoption of a stabilization program in its first year. Nzé believed that the Intermediate Plan, in compensation, "benefitted from the lessons drawn from the preceding plans," in so much as it limited forecast spending even though oil revenues were high. Similarly Nzé believed the Fifth Plan could increase public spending even though, in the oil price crash of the 1980s, oil revenues were comparatively low.[16]

The key phrase for the Fifth Plan was one referring to an inevitable future which the planners called the "*après-pétrole*" (post-petroleum era) and which referred to the state of the economy once oil reserves were exhausted. According to the logic of the *après-pétrole*, it was essential that the government develop diversified sources of economic growth. Principal among these was agriculture. The key phrase was "auto-développement"—a dynamic process by which young farmers would expand state-funded projects on their own initiative (with government support) and revitalize the rural sector (p. 57).

The main innovation in the Fifth Plan's rural development program was the way in which it approached its objectives. The objectives of rural development themselves had changed little from the previous plans: to achieve self-sufficiency in food production; to reduce the rural exodus; to increase cash crop exports; to ameliorate rural conditions; etc. (p. 57). But where the Fifth Plan had learned from its predecessors was in the manner that it transposed ends and means in rural development. In the previous plans agricultural production was approached as an end in itself—a measurement of the government's commitment to its rural populations. In the Fifth Plan agricultural production was approached as a means to meeting self-sufficiency. "This global objective ought to be interpreted in terms of reducing the food dependency since, save for a radical modification in eating habits the country is not, ecologically speaking, capable of producing all its food needs" (p. 105).

The Fifth Plan differentiated itself from its predecessors by taking a complete inventory of food consumption as its starting point, and then forecasting the amount of acreage required for new planting in order to meet future consumption needs. Banana consumption in 1982, for example, was 149,000 tons. Forecasted banana consumption in 1988 was set at 161,000 tons. At an estimated production of 15 tons/hectare this corresponded to 1,000 hectares commitment on the part of the government for new plantings. Manioc and taro tubers were by such calculations deemed 750 hectares deficient for 1988 projected consumption (p. 106).

When it came to meeting consumption requirements for such cereals as rice, corn, and soy, the government factored in import levels, so that rice consumption in 1983, for example, of 14,000 tons was identified as being met by 13,600 tons of imports. With a forecast consumption of

19,700 tons of rice by 1988, the government estimated a foreseeable deficit of 19,300 tons (given that national production of rice was a paltry 400 tons). The production objectives in the Fifth Plan were for 5,000 tons of rice, leaving an estimated 14,300 tons of imports necessary to meet projected consumption (pp. 108-109). Corn imports were to be totally eliminated by adding 1,430 hectares of surface area for corn production; soy imports by adding 3,900 hectares for its production (pp. 108-109).

For the first time, food-crop production was being planned with its ultimate *demand* in mind. Previously it had been planned as a sort of rural welfare, a subsidized activity to keep peasants employed. Now it was *a means to some greater purpose of its own*: i.e., rural production was meant for Gabonese *consumption*.

Consumption of meat and other animal products was calculated and forecast based on needs projected. Cattle used for meat were consumed in great amounts, 12,600 tons in 1982, out of which the national production of 100 tons met only .8%. The government planned to increase production to 3,400 tons, or 28.5%. Sheep and goats were forecast at 2,200 tons consumed by 1993, and 3,300 tons by the year 2000, when the government hoped to be producing 1,600 tons nationally, or 48.5% of its needs. By the year 2000, the government planned to be meeting 83% of its pork, 89% of its fowl, and 100% of its egg consumption (p. 111).

Another lesson learned from the mistakes of the past was that cash crops and food crops should be integrated in their development, and that they should be diversified within every region so as to overcome the problems associated with poor infrastructure and internal communication. To consolidate past accomplishments the planners established *Opérations Zonales Intégrées* (OZI). An OZI was geographically delineated "by a zone surrounding an already existing pole of development" and contained a *centre d'appui technique*, various cash crop projects, and cultivation of food crops. It was supposed to be managed in a decentralized and autonomous manner, as much as possible, and according to the priciples of profitability (p. 59).

In the Ogooué-Lolo region there were two OZIs: The

N'Djolé OZI produced cocoa and coffee as export crops, and *cultures vivrières* on a surface area of 250 hectares in 1988; the Lastourville OZI produced similar crops on around 500 hectares. In the Ogooué-Maritime province the Fernan-Vaz OZI contained a coconut plantation and also grew plantain bananas and other garden variety vegetables, as well as possessing some commercial fishing. In the Woleu N'Tem region there were four OZIs: The Oyem OZI grew cocoa and coffee, vegetables, on 1,250 ha. along with ranching and fish-harvesting facilities; the Bitam OZI grew hévéa and cocoa and food crops, with some fishing, on around 3,000 ha.; the Mitzic OZI grew hévéa and vegetables, and had fish harvesting on 500 ha.; and the Médouneu OZI was preparing to grow market vegetables (pp. 118–119).

In the Estuary region two OZIs were identified: The Kango OZI produced hévéa and cocoa on around 2,000 ha. and was set to grow vegetables, to ranch, and to contain both commercial fishing enterprises and fish-breeding projects; while the N'Toum OZI at the village Akok was to continue growing rice as it always had, but additionally was earmarked for a market vegetable project, and a pork ranch, and a fish-breeding center. In the Haut-Ogooué region the Okondja OZI grew coffee and cocoa on 150 ha. of land, and was going to expand to around 4,000 ha. to add ranching, fishing, and a market vegetable project. In the same region two other OZIs, at Leconi and at Franceville, were going to ranch, grow market vegetables, and raise fish. In the Moyen-Ogooué region the Lambaréné OZI raised cocoa and had a palm oil plantation on 1,550 ha., and was going to expand an additional 1,000 ha. to fit a ranch, a fishery, and to grow various foodcrops. In the region of N'Gounié the Mouila OZI at Mimongo grew coffee and cocoa on 250 ha. of land, and was going to double that surface area for ranching and foodcrops. Meanwhile the N'Dende OZI was also planned in that region for rice, fisheries, fruits, and vegetables. In the Nyanga region the Mayamba OZI grew hevea and coconuts on 3,200 ha., and was planned for fruits and vegetables and a fishery. The Moabi OZI in the same region grew palm and was designed to grow plaintain bananas and some other market variety vegetables. In the Ogooué-Ivindo region, OZI I and OZI II were planned for cocoa and coffee and *cultures vivrières,* and Koumaméyong OZI for hévéa, on around 3,000 ha. of land (p. 119).

Financing the Fifth Plan's OZI program was not going to be easy, as it was estimated to cost around 64.3 billion CFA francs over the 1984-1988 time period out of a total of 122.7 billion CFA francs for rural development expenditures (pp. 129-130). But as Howard Schissel observes, in one of a series of articles written on Gabon in the early-to-mid-1980s: "[T]he soaring value of the American dollar, the currency in which almost 90 percent of Gabon's exports are denominated, has assured Gabon a high level of revenues."[17]

This bright and optimistic situation was however dimmed by the precipitous drop in the price of oil during the decade known today as the "oil price crash." By the spring of 1986 the price of a barrel of crude oil on the spot markets had fallen below the $10 per barrel marker and drove government revenues down to new lows. As could be expected, the government announced severe budget cuts (50 billion CFA francs in 1986 and 90 billion CFA francs in 1987).[18] In December of 1986 the new Finance Minister, Jean-Pierre Lemboumba Lépandou, met with private creditors in Paris for negotiations on Gabon's debt (which had risen to 400 billion CFA francs) and soon was working on the details of yet another IMF-imposed austerity plan.[19] The Paris club meetings in January of the following year effectively derailed most of the new commitment to rural development projects outlined in the Fifth Plan as debt repayment, combined with recurrent expenditures, forced a reduction in the investment budget to 100 billion CFA francs.[20]

As the country's economic crisis worsened, money set aside for development projects continued to dwindle. *By the end of the period of the Fifth Plan few of the objectives set out for rural development had been achieved, and agriculture was producing less of the gross national product than ever before.* The era of five-year development plans had finally come to an end. It was understood that under the great uncertainty of petroleum prices there was no way for the government to guarantee five-year commitments to its development budgets. Instead the Ministry of Planning introduced a new two-year plan, which removed the word *"développement"* altogether from its title.

The government would pursue its developmental objectives under the *Loi-Programme 1989–1991 des Investissements Publics et Parapublics (Loi 24/88* December 31st, 1988) which was in turn updated in December 1989 as the *Loi-Programme 1990–1992 des Investissements Publics et Parapublics (Loi no. 14/89,* December 31st, 1989).

According to the new logic of investment, which had by this time begun to replace the discourse of development, the "priority of productive investment" was to be for "the continuation of projects already in effect."[21] The program maintained three of the long-standing objectives (amelioration of rural lifestyle, increased profits and rents from agriculture, self-sufficiency in foodcrop production) but adopted a new liberalized strategy for realization of these objectives that focused on "the optimization of costs" and the "potentials for development" of each project (p.40).

By the 1990s, the entire discourse of development had changed to a new emphasis on profitability and economic rationale. The government established three economic criteria by which rural development programs were to be evaluated. The first of these was that the pursuit of any project would depend on whether or not its "economic and social cost was higher than that of its pursuit" (p.41). Programs would no longer be funded simply because they met certain objectives defined by the government planners, but they would be suspended or abandoned if the costs were deemed too high. The second criterion required that for any new project to be launched it would have to "demonstrate economic and financial profitability" (p. 41). Public revenues would no longer flow into projects unless those projects could show a return on investment. The third criterion was that choices for programs would be made with "an optimal utilization of resources already engaged" kept in mind in order to promote a "dynamic development of the zones in which they were implanted" (p. 41). Allocation would therefore no longer be disjointed, but would attempt to synergize with other expenditures.

These criteria were surely the result of IMF and World Bank influence, but they nevertheless became the language of government policy in Gabon. After a quarter century of trying to spread public funding for rural and agricultural development evenly throughout the territory, now the government was making the hard choices, selecting the winners from the losers. The profit motive may not have been new to Gabon, but its application to rural development projects was.

Out of all the ambitious plans for rural development that the government had pursued in the past, the operations that were to receive fresh public funds were divided into six major programs: the *hévéa* program, the palm oil program, the cattle ranching program, the cocoa and coffee program, the peasant supervisory program, and the accompanying research programs. Altogether the government planned on investing 56,189,000,000 CFA francs into agricultural development over the course of 1990-1992.[22] In preparation for the *après-pétrole*, the government treated these expenditures as investments—that is, as money spent with the expectation that it would produce a return.

The days of rural "welfare" were over. Farming and ranching were to be treated like any other business. The hevea project at Mitzic, for example, was to receive the largest infusion of new public revenues because the trees planted there had performed better than at Bitam, Kango, or Mayumba (p. 43). The emphasis on Cattle ranching over other breeds reflected the high price of beef in Gabon and therefore cattle ranches' relative profitability (p.45). The new cocoa and coffee plantations were earmarked for development because they had attracted private investors: e.g., Banque Islamique de Développement, Banque Arabe pour le Développement Economique en Afrique (p. 46). And out of the plethora of OZI identified in the Fifth Plan, only six, the most optimally located, were to receive new allocations from the government (p. 48).

Perhaps such rationales must be treated as justificatory rather than explanatory. The goverment had to justify new investments in a climate of fiscal austerity and decreasing state revenues. But the purpose of investing was at its most basic remarkably different from the past purposes of allocation.

Ultimately the lesson to be learned from all of the previous development plans was that *spending is not enough.* In the 1960s the government believed that through its rural development expenditures it might curb the rural exodus that was draining its countryside and flooding its cities. By the 1990s, however, more than half the population lived in the three major cities: Libreville, Port-Gentil, and Franceville.[23]

In the 1960s the government hoped that by experimentation, popularization, and commercialization of cash crops it could increase the rural income levels and thereby reduce the state of economic dualism that existed between town and country. By the 1990s however—after a quarter century of plantations and pet projects—agriculture accounted for only 5.8% of total exports (1990) or *around a third of what it had represented before the government's first development plan.*[24]

In the 1960s the government felt that through spending money on the promotion of village life it could minimally provide for a farming base capable of food self-sufficiency. By the 1990s, however, Gabon was almost completely dependent on food imports, which consumed almost a quarter of the total import bill.[25]

The illusion of allocation *as* rural development could not withstand the facts.

As Gabon would eventually need to survive on its development projects (rather than the other way around) the old idea that rural development could be purchased like a commodity dissolved. Rural development is not something that can be bought with money, but rather, is a process. For better of for worse, the equilibrium between the traditional and the modern sectors would have to occur somewhere outside the domain of public finance.

NOTES

1.   *Economie et Plan de Développement: République Gabonaise* (Paris: Ministère de la Coopération, 1963), p. 14.

2. Ministère de l'Economie Nationale, Le Commissariat au Plan, *Premier Plan de Développement Economique et Social de la République Gabonaise* (Libreville, 1966), p. 157.
3. *Deuxième Plan de Développement Economique et Social de la République Gabonaise* (Libreville: Ministère du Plan, 1971), préface.
4. *Troisième Plan de Développement Economique et Social de la République Gabonaise* (Libreville: Ministère du Plan, 1975), p. 83.
5. *West Africa*, (February 17, 1975):203.
6. *West Africa* (June 23, 1975):731.
7. *West Africa* (March 17, 1975):203.
8. *West Africa* (May 30, 1977):1025.
9. *Troisième Plan, op. cit.*, p. 88.
10. *West Africa* (August 15, 1977):1705.
11. *West Africa* (July 14, 1980):1313.
12. *West Africa* (August 18, 1980):1537.
13. *Plan Intérimaire de Développement Economique et Social de la Republique Gabonaise 1980-1982* (Ministère du Plan, du Développement et des Participations, 1979), p. 30.
14. "Gabon's Rural Waste: Eric Rondos Reports on the Neglect of Gabon's Agricultural Sector," *West Africa* (August 18, 1980):1537.
15. *Plan Intérimaire, op. cit.*, p. 56.
16. *Cinquiéme Plan de Développement Economique et Social de la République Gabonaise* (Libreville: Ministère de la Planification, 1984), p. 7.
17. Schissel, "The ups and downs of an open economy," *West Africa* (October 15, 1984).
18. *West Africa* (February 17, 1986):385.
19. *West Africa* (December 22/29, 1986):2675.
20. *West Africa* (January 12, 1987):92.
21. *Loi-Programme 1990-1992 des Investissements Publics et Parapublics* (Libreville: Présidence de la République, 1989), p. 4.
22. Ibid., *Annexe, A3.*
23. Hugues Barro Chambrier, *L'Economie du Gabon: Analyse Politique d'Ajustement et d'Adaptation* (Paris: Economica, 1990), p. 22.
24. Loi-Programme 1990-1992, *op. cit.*, p. 27.
25. Chambrier, *op. cit.*, p. 430.

Chapter 5

The Transgabonese Railroad and the Modern Sector

ALLOCATION AS/AND MODERNIZATION AND DEVELOPMENT

Rich people often think they can buy anything, and that everything has a price. Rich states often fall into the same trap. This is what happened to Gabon during the oil boom of the 1970s. Omar Bongo believed that he could purchase modernity for his country and, in so doing, make Gabon more like Europe. The fallacy in this logic is that modernization is not a commodity, but a process. As such one can neither buy nor sell it. This does not mean that European and American businessmen weren't willing to allow the Gabonese to operate under this illusion. After all, if you're selling technology, why not package it as modernization and development? The theme of this chapter is that, despite hundreds of billions of CFA francs in allocations to the so-called "modern" sector, Gabon has not modernized. In fact, these allocations may have retarded the process of genuine modernization and development in Gabon. For the

Gabonese planners have treated allocation *as* development and, as such, have fallen into the trap of the rentier.

THE TRANSGABONESE RAILROAD

If one project epitomizes the government's effort to be a
so-called "modern" country, then it would have to be the
transgabonese railroad (referred to hereinafter as "the
Transgabonais"), a project started in 1972 and not completed until 14 years later, at a total estimated cost of
around 700 billion CFA francs. This is an ideal subject with
which to start a discussion of the planned development of
Gabon's "modern" sector. First of all, because they said it
could never be done. Even during the twilight of French
colonial rule, when all the resources of the mother country could have been marshalled for its construction (as
they had been for the Congo-Ocean railway), the engineers
under then-governor Emile Gentil refuted the feasibility of
a railroad cutting through the Gabonese rainforest. Later
on, when Gabon asked the World Bank for funding to build
its railroad, and was refused—on the grounds that it was
too expensive given the limited resources of the country,
and that it would never be economically viable given its
tremendous costs—president Bongo was reported to have
said: "If it is necessary to strike a bargain with the devil to
build the railway, I am fully ready to make such an arrangement."[1]

The fact that the railroad ever was finished is no small
miracle, and an accomplishment of which most Gabonese
are purportedly quite proud. However, we start our discussion of development of the "modern" sector with the
construction of the Transgabonais because it demonstrates that—given enough money—no project is too
expensive nor too wasteful to stop a determined rentier
state from developing it.

On an ideological level the railroad symbolized Gabon's
deep and enduring commitment to the capitalist path of
modernization and development. In contrast to the rural
development plans, which were promoted by the
Gabonese government through public spending and cen-

tralized planning (more commonly associated with a state socialist tradition), when it came to modern, industrial-looking projects: "the private sector—and it alone—is capable of providing the capital, the manpower, the experience, and the opportunities without which no industry could function."[2] This philosophy, enunciated in the First Plan, would be echoed in later efforts at building a modern infrastructure and industrial base. Private French, German, Italian, Belgian, Dutch, and English firms: These were the principal actors in the EUROTRAG consortium that constructed the Transgabonais.

Finally, we begin the discussion of the modern sector in Gabon with the Transgabonais because it confirms, in a very material sense, that the Gabonese government intended to define its economic modernization as the expansion of its primary extractive industries as much as (if not more than) its manufacturing base. The railroad would open up previously unexploited rainforests, permit the trans-shipment of manganese and uranium across national territory, and make accessible the vast quantities of iron ore that were located in the impenetrable northwest corridor of the country. "The presence of the railroad" claims the First Plan, "will be followed by a remarkable augmentation in the gross domestic product" (p.91)

So great were the government's expectations that in its First Plan there were two sets of figures provided for future economic growth: one *with* and one *without* the railroad. "The influence of the complex," the planners explain, "will be such that we must envision two eventualities that might present themselves" (p. 72). For example, the planners claimed that without the railroad the gross domestic product of Gabon would only be 74,014 million CFA francs in 1975, and 77,155 million CFA francs in 1980; while *with* the railroad the gross domestic product would be 96,827 million CFA francs in 1975, and 117,826 million CFA francs in 1980 (p. 91). In every sector of the Gabonese economy, from forestry to mining to agriculture, the railroad was predicted to contribute to increased output. Is it any wonder that president Bongo was willing to, in his own words, "strike a bargain with the devil" in order to build it?

In his prefatory remarks to the Second Plan, President Bongo writes that the state would "concentrate its efforts on the railroad, which is the priority project around which our development will support itself."[3] His statisticians and civil engineers had estimated its costs to be around 30.5 billion CFA francs for the first leg of the railroad, from the coastal port of Owendo near Libreville to Booué in the center of the country. He optimistically expected 10 billion CFA francs in grants and aid from the French, and around 20 billion CFA francs in loans from a variety of sources including the World Bank (pp. 87-88). Moreover his Second Plan estimates that the government could/would repay these loans by 1977 (p. 88).

It is of more than passing interest to observe that the government expected to turn a profit on the Transgabonais. Economic rationale for projects of a modern industrial nature such as the railroad were expected to meet criteria of "profitability" in a way not expected of equivalent traditional agricultural projects. Because it would open up essentially unexploited forest reserves to commercialization, the Owendo-Booué leg of the railroad was expected to be profitable from the traffic in wood alone (p.85). Great effort is made in the Second Plan to show/calculate the profitability of the railroad. Why was this done here, but not in the plan's section on rural development?

One reason may be that the development plan, like other official government documents that discussed the railroad, was in part an advertisement for the Transgabonais. Since the railroad was in need of substantial external funding, especially loans, it had to appear to be profitable in a way that rural development projects, which were seen as state welfare programs, did not.

Another reason, perhaps more subtle, was that the government itself believed in the profitability of modern industrial projects in a way that it did not for farming and ranching and such. As shall be seen, the country was essentially preindustrial at the time of the First and Second Plans. The promise of modernity may have carried with it certain hidden assumptions about technologically advanced projects which had been denied Gabon for so long. But if a criterion of *rentabilité* [i.e., "profitability"] was established for these projects, as it would eventually be established for rural development ones, can we assume that the industrial sector would succeed where the agricultural sector failed? Was the profit motive really that important after all in determining the success of a project?

The successful completion of the Transgabonais railroad was accomplished much more by force of political will than by its own economic attraction. To start with the most immediate challenge to its profitability, the World Bank outright rejected the plan for the railroad as economically nonviable. (The World Bank felt that the existing transportation system could export Gabonese minerals without the need for a new railroad, and that the cost of building a train through the middle of Africa's equatorial rainforest was going to be too expensive to ever pay off.) This rejection by the World Bank forced president Bongo to actively seek out other investors for his pet project. But without the backing of the World Bank it was difficult to convince potential investors that the railroad would in fact be profitable (and by logic, a good investment). The World Bank's veto was in fact a reversal of its own previous position. In the late 1960s the World Bank had backed the idea of a train linking Gabon's mineral-rich interior regions with the coast, but in February of 1973 refuted this view on the basis that the future of iron ore on the world markets was uncertain, and revenues from forestry were unlikely to be sufficient by themselves to justify the tremendous fixed costs.[4]

President Bongo defiantly announced that "[t]he Transgabonais Railway will be built. It will be built by one means or another with the help of this country or that country...."[5] His trips to Libya and other North African Arab states in the Fall of 1973 were very much efforts to muster financial backing from these countries for his railroad. A year later, when support was not forthcoming, he was quoted in *Le Monde* as saying, rather bitterly, that he "was not satisfied with the state of relations between his country and the Arab states."[6] Embittered by Colonel Khaddaffi's apparent reversal on promises of financial support for Bongo's railroad project, the Gabonese president undiplomatically remarked that "[t]hese are people who are hardly serious and who do not bother to abide by the agreements they have signed."[7]

Eventually, he was forced to rely on his country's traditional source of aid—France.

In resorting to French assistance, president Bongo surely had struck his bargain with the devil. French Minister of Finances, Valéry Giscard d'Estaing, was on hand in Libreville for the ceremonies that officially started work on the first leg of the project. December 30th, 1973, was also president Bongo's birthday, thereby linking the birth of the railroad with that of the man who had engendered it. But French firms held a majority (50.7 percent) interest in the EUROTRAG consortium, the firm that had been contracted to build the Transgabonais. Therefore operational control did not belong to Gabon, but to France.

So what? Almost a decade later certain anonymous government officials were reported to have said that cost overruns were purposely built into the construction of the railroad to "pad profit margins" of the French-dominated consortium.[8] In his rush to build the railroad, and in his strident advocacy of the private sector's privileged role in modernization, president Bongo had apparently been blinded to the darker aspects of the profit motive: i.e., corruption and greed.

Perhaps more than any other single factor, even the political will of the president himself, what ultimately made the construction of the Transgabonais railroad possible was the 1973 oil crisis. Political will and promises of big profits are, after all, not enough. It takes money to build a railroad. What the first oil shock provided to the state was not only a sudden and massive influx of external rents generated by the petroleum sector, but even more important, the credit worthiness it needed to borrow the money required to continue the project that it had started.

The construction of the first leg of the railroad experienced considerable cost overruns and delays. To meet the growing expenses the government earmarked extra funds in the 1975 budget—which had *tripled* government spending in one year—for its railroad. President Bongo said in an interview in the Fall of 1974 when the draft budget was being approved: "I want to add that the spectacular increase in our budget revenue should enable us to realize the basic infrastructures which we lack."[9] A major share of the development money identified in the budget (2,300,000,000 CFA francs) would go to the Transgabonais.

In addition to increased allocation of oil rents, the government was able to borrow substantial amounts of money for the project. For example, in the Summer of 1975, the Gabonese Finance Minister visited Morocco and signed an agreement under which Morocco would lend Gabon $28 million for financing part of the Transgabonais.[10] At the same time Gabon became the first black African country to raise money on the Eurobond market, offering a 10-percent coupon on a $60-million issue for five years to help finance the country's development plans.[11] President Bongo meanwhile visited Japan and received a loan for 3 billion yen to purchase rolling stock and other equipment for the railroad.[12] In the summer of 1977, the Office du Chemin de Fer Transgabonais (OCTRA) borrowed $10 million to finance important civil engineering works on either side of the Junkville tunnel which formed part of the Transgabonais railroad construction.[13]

Money was not a concern when it came to the Transgabonais and so it is no surprise to discover that even when the country was in the midst of its 1977 IMF austerity plan, money was still to be found for the railroad. Cost overruns were beginning to reach alarming proportions. Nevertheless president Bongo was able to secure a 29.5-million-mark ($13.4-million) loan from West Germany.[14] And on a visit to Riyadh he secured another $20-million loan from the Saudi Arabian Development Fund.[15] By 1979, when the first microsection of track running 183 kilometers from Owendo to N'Djolé was completed, the project had consumed more than $1 billion—or *twice the original estimate.*[16]

Speaking from the platform at N'Djolé station, alongside French minister of the economy René Monory, president Bongo recalled how "France was the first country to offer cash aid to start construction" and hoped the Transgabonais would "consolidate our national unity."[17] It was during this speech that president Bongo also made clear his government's intentions to continue the railroad

construction all the way to Franceville. This is an important shift from the original plan, which was to connect
Owendo with the iron ore deposits at Belinga. When the
World Bank had originally rejected the Transgabonais, it
was based on the prospect that iron exports would be
transported from the interior to the coast. Now president
Bongo wanted to extend a branch to his home town as the
second stage, with the only economic rationale being the
transshipment of manganese and uranium (already being
exported by other means).

Although the World Bank would probably have been
even more querulous at the economic justification of the
Booué-Franceville extension (were it consulted, which this
time it was not), the second leg was politically very important to president Bongo. Not only did it open up new territory, but (as French journalist Pierre Péan has suggested)
it permitted president Bongo to make a play for tribal chief-
tainship of his Batéké people. Michael Reed has added, in
his discussion of the project, that president Bongo eventually ordered an extension from Franceville to his rural
birthplace 40 kilometers away in order "to demonstrate to
his own tribespeople that he had done something for
them."[18]

The total cost from Owendo to Franceville was estimated to be somewhere around 48 billion CFA francs,
according to the Intermediate Plan.[19] But even at the time
planners admitted that "there is no way of knowing that it
might eventually not be more, in view of the considerable
obstacles to construction."[20]

In the Spring of 1979, Gabon was able to raise $70 mil-
lion on the Eurocurrency markets for the Transgabonais'
second leg.[21] By 1981 the government had budgeted nearly
80 million British pounds equivalent for the project, and
had commenced with work on the Booué-Franceville
branch even before the Owendo-Booué leg had been completed. On a return trip from a meeting with European ministers in January of 1982, president Bongo announced that
the Belinga branch would not be financed at all, and was
subsequently abandoned.[22]

In addition to the money it borrowed, the state also continued to spend public revenues—*principally generated by*

the petroleum sector—on the completion of the Transgabonais. In Article 4 of Loi no. 1984, which officially allocated funds for the Fifth Plan, we find that more money was going to be spent on construction of the railroad than on any other developmental project or social service. The railroad was earmarked to received 100 billion CFA francs in 1984, 102 billion CFA francs in 1985, 95 billion CFA francs in 1986, 46 billion CFA francs in 1987, and 18 billion CFA francs in 1988—for a grand total of 361 billion CFA francs. Now compare this figure with that of the 122 billion CFA francs committed to agriculture over the same five-year period; or to the 212 billion CFA francs to be spent on education and social services.[23] Estimations of cost that factored in a hypothetical inflation rate were included in the Fifth Plan and showed a real total cost of 408 billion "current" CFA francs (p. 216). When the Transgabonais was actually completed the cost of the second leg ran up to 611.6 billion CFA francs.[24]

Georges Rawiri, the Minister of Transportation at the time of the railway's completion, was reported to have said that "forgetting construction costs, there is no possibility of the railway making an operating profit."[25]

In all, the construction of the Transgabonais required the felling of six million trees, the building of fifty bridges, the digging out of one tunnel, the employment of 4,000 workers (around half of them Gabonese nationals), and the allocation of somewhere on the order of $4 billion.[26]

French participation remained an essential ingredient in the recipe for the railroad's completion. Two British firms had been able to purchase 11% of the EUROTRAG consortium's interests each, reducing the French share to 39.5%, while leaving the remaining partners unchanged (West Germany 16.8%, Italy 12.1%, Belgium 4.8%, and Holland 4.8%)[27]. Despite the entrance of the British, France's role remained central. At the official inauguration ceremonies held on December 30th, 1986 (the president's 51st birthday) both the French Prime Minister, Jacques Chirac, and the president's son, Jean-Christophe Mitterrand, who served as special adviser to his father, were on hand to commemorate what represented both an achievement of grand proportions for Omar Bongo, and a profitable venture for the French.

Yet the project was anything but profitable for Gabon, at least in the short term. The railway almost immediately required a state subsidy of $60 million a year to finance its timber and passenger traffic. The railway authorities had to suspend first-class service for its passengers, and had to hunt down adequate staff and payroll to maintain its existing rolling stock. Iron ore, of course, was not transported on the railroad, since the Booue-Belinga branch had been suspended in order to connect the port of Owendo with president Bongo's hometown. Meanwhile the manganese that was supposed to now be transshiped via the railroad could not because the port at Owendo was incapable of storing it. To make matters worse "[t]he railway's finishing touches will use half the CFA 100 bn set aside for investment in the 1987 budget."[28]

Not only was the railroad unprofitable, but it was costly. According to the World Bank, the Transgabonais was "symptomatic of the inefficient use of public resources" by Gabon, consuming "more than six times the comparable international standard cost."[29] Whether or not this figure is accurate (and given the World Bank's disapproval of the project from the outset one might assume that it exaggerates), the burden of building the railroad was a heavy one. For example, Hughes Alexandre Barro Chambrier isolated that part of the development budgets devoted to the construction of the project and came up with the following figures: 16% (1971), 34% (1972), 10% (1973), 6% (1974), 12% (1975), 9% (1976), 24% (1977), 27% (1978), 23% (1979), 25% (1980), 29% (1981), 25.7% (1982), 30% (1983), 39% (1984), and 34% (1985).[30] Even this kind of distortion was not enough to adequately finance the project.

Borrowing heavily to meet the inflated costs, the state increased its total public debt from 19.8 billion CFA francs in 1974 to 147 billion CFA francs in 1986.[31] Debt service payments soon were consuming over a third of budgetary receipts, suggesting that the total real cost of building the Transgabonais railroad was very dear indeed. After all, how much of the total capital available to government ought to be spent on one project? At what price does modernization—for that is what the train symbolized—come?

President Bongo might answer such questions in the same way he responded to other queries into his country's modernization and development plans: "We want to do in a few decades what others took centuries to do."[32]

It is in the nature of rentier states to seek shortcuts to economic development, to try to buy with oil money what is lacking in the socioeconomic system. By spending hundreds of billions of CFA francs on the Transgabonais, president Bongo hoped to rectify centuries of infrastructural neglect and, more important, created the *illusion of progress*. After all, wasn't it a great symbol of progress in the United States during its own nineteenth-century era of economic development to link the East coast with the West via the Transcontinental railroad? Can we ignore the massive symbolic meaning of a commensurable railway crossing the heart of Gabon? Surely the successful completion of the Transgabonais has been a symbolic and political triumph for president Bongo and his regime. But in economic terms any such comparison with the American Transcontinental railroad falls short. For there were no pioneers heading to the interior of Gabon to carve out a new future for the country (unless you count French logging companies). There were no Gabonese robber barons to profit from its business and form the basis of its indigenous capitalist class (unless you count corrupt French businessmen and a few friends and family members of president Bongo). The Transgabonais, metaphorically speaking, is a train that leads nowhere. It is, so to speak, an iron path to more rentierism and primary extraction.

THE INDUSTRIAL SECTOR

Now we turn to the talk of industrialization in Gabon, which, like the story of the Transgabonais, speaks to the problem of a country that wants to purchase development. "Our economy is liberal" claimed president Bongo in an interview during the heyday of the oil boom: "[T]hat is, it endorses free enterprise, but in a framework of planning under state authority."[33]

President Bongo liked to call his economic philosophy "planned liberalism" and believed that it was the only way for his country to do in a few decades what others took centuries to do. In his mind the only actors that were capable of creating industry were private corporations, whom he courted "liberally." Yet the history of the Concessions Regime in 1898 proved that the danger of letting foreign firms run wild in the country was their degeneration into a psychosis of feral capitalism.

When it came to manufacturing, Gabon was the kind of country that one could quite justifiably call "pre-industrial." Looking at the *Economie et Plan de Développement*, we find that the entire manufacturing base of Gabon at independence consisted of a little more than 15 sawmills, four wood factories, 15 coffee mills, four palm oileries, one rice processor, and one soap factory.[34] And most of these had been created only recently under the assistance of FIDES after the Second World War. As the French planners write, "[i]ndustrialization in Gabon hasn't even begun" (p. 22). There was little or no real "indigenous" capitalist class separate from the ever-dominant French businessmen. Under such circumstances the Gabonese state had to act in a central and purposeful way to achieve real industrial growth. Yet it lacked the financial means to do so.

The industrial development strategy of the First Plan was designed to favor industries that turned raw materials into semi-raw commodities (*industries de transformation*). This strategy was adopted by the government because the so-called "handicaps" that were identified in the First Plan—limited and low-skilled workforce, absence of an entrepreneurial spirit, geographical isolation, and a small domestic market—were only offset by two so-called "advantages": The first of these was a "variety and abundance of primary materials" and the second was the "possibility of producing cheap energy" for their transformation.[35] Subsequently, the industrial projects identified in the First Plan used the country's natural resources and reflected the economy's primary extraction sector: a cellulose factory, a palm oilery, a petroleum refinery, a cement factory, and so on (pp. 303–307).

The Gabonese industrial strategy also rejected from the outset any idea of establishing state industries. "The state cannot do everything," warns the First Plan. "We feel that in the new, young, and consequently inexperienced countries, the recourse to state industrialization ... is hazardous" (pp. 291-292). Distancing itself from an African state socialist path of industrial development, the Gabonese economic system was called by some observers a model of "African Capitalism."[36] Léon M'Ba did not advocate state industries, but rather endorsed and invited private ones to exploit his country's vast natural resources.

Before looking individually at the industrial companies, it is important to note that the Gabonese state did take an active ownership in all supposedly private enterprises. This occured for two reasons. First, state participation was required for those industrial projects which were unlikely to have received sufficient funding without state promotion because they were by nature in the interests of the state, rather than profitable ventures in their own right. Second, the state was able to acquire participation in those projects which were likely to turn a profit because of the application process which required any firm entering the country to offer a share of its capital stock to the state for purchase, when the state exercised its option.

Some examples will illustrate these points. Despite the rethoric against state-owned industrial enterprises, for example, we find that the government of Gabon had purchased 45% of the capital of the Société Gabonaise de Cellulose (SOGACEL). This company had been first envisioned by the government in 1963 as a way to create a national industry that would use the country's comparative advantage (i.e., abundant rainforests) to export cellulose to the world markets. Also involved in the company were: Centre Technique Forestier Tropical (10%); Franco-American corporations consisting of Lilles, Bonnières, Colombes, and Allen (35%); and Banque Internationale pour l'Afrique Occidentale et Crédit Commercial de France (10%).[37] Since the ultimate interests to be served by the cellulose factory were national interests, in this case the state participation was required for the project to survive.

By way of contrast, the Gabonese state was only able to acquire a 5% share of the capital of the Société Equatoriale de Raffinage (SER) which was the company responsible for the oil refinery at Port-Gentil. The oil refinery was a project that originally came about when the heads of state from the five UDEAC countries met to discuss the creation of a refinery that could meet their regional petroleum needs. Applications were tendered by the Congo for Pointe-Noire, by Cameroun for Douala, and by Gabon for Port-Gentil. Based on a study conducted under the auspices of the French Bureau de Recherches du Pétrole (BRP) and the Compagnie Française des Pétroles (CFP), the Port-Gentil site was ultimately selected and Gabon gained what was perhaps the single most important industrial project of its history. The SER was founded on October 7th, 1964, with the distribution of its capital divided among the following participants: 27.5% for BRP, 37.5% for CFP, and 5% for each of the five UDEAC member states.[38] Since the primary interests to be served by the oil refinery were regional rather than national, Gabonese state participation was not essential to the project's survival, but was rather a product of its territorial sovereignty and its membership in the customs union. Still, the *bilan* of this project was largely positive for the Gabonese economy, as the authors of the First Plan were ready to admit, because "the refinery could be the beginning of a much vaster industrial complex and notably favourable to the creation of a chemical industry" (p. 306)

There were other, smaller-scale projects designed for the domestic markets. Examples of these import-substitution projects in the First Plan were a brewery, a flour mill, a biscuit factory, a confectionery, a fish smokier, a panty hose factory, a paint factory, a metal shop, and a furniture factory—all of which were to be entirely financed by private funds (pp. 310-318). Most of these businesses were going to employ between 20 and 40 salaried workers. "The total number of employees created in the industrial manufacturing sector," concludes the First Plan, "will be around 830" (p. 318). That number includes the estimated employment figures of the refinery and the cement factory, which by themselves would account for more than a third of these new industrial jobs.

Even for a country the size of Gabon, with a population of feeble proportions, 830 jobs was not the stuff of industrial revolution. But if the number of industrial jobs created, or even the low 8.2% of total development expenditures dedicated to industrial investments in the First Plan appear small, the First Plan defends itself with the bromide: "Industrialization is not an end in itself" (p. 292). The government's approach to projects of an industrial nature was profit-oriented from the outset. And if the balance between profits expected and jobs created were to tip in favor of the former over the latter, then that was acceptable to the state. Besides, as the First Plan admits: "[T]he managers, supervisors, and a good part of the highly qualified workers will be recruited from outside Gabon" (p. 318).

The First Plan reflected attitudes toward industrialization held by members of the Gabonese government under the presidency of Léon M'Ba. Part of his government's hesitancy to become more involved in rapid industrialization may have derived from M'Ba's dependent relationship to France. *French business, it must be remembered, virtually controlled the Gabonese industrial base.* There were deeply entrenched French business interests in Gabon that were hostile to import substitution of French goods, unless they themselves could be manufacturers. This may explain why all the industrial programs designed to manufacture for the domestic markets were "private" sector businesses. So too the export-promotion industries derived from the production-and-extraction endeavors were controlled by French interests.

On another level, the attitude of President M'Ba during his last years of office (when the First Plan was unveiled) to the new urban working classes could be described as suspicious or even fearful. According to figures produced by the International Labor Organization, the number and intensity of industrial disputes had increased ferociously. In 1960 there were 31 industrial disputes involving 1,522 workers and resulting in a total of 2,205 working days lost. In 1962 there were 36 industrial disputes involving 2,374 workers and resulting in 4,028 workdays lost. In 1964 there were 51 industrial disputes involving 4,694 workers and 12,388 workdays lost.[39] After the attempted coup of 1964, unions were banned, and industrial disputes slowly disappeared.

Despite the suppression of union activities that occured during its implementation, the Second Plan

(1971–1976), brainchild of President Bongo, reflected a changed attitude on the part of the government toward the industrial cadres. It recognized the need for a better trained workforce, and was the first plan to recommend the establishment of technical schools and professional institutions in Gabon to "satisfy the needs of the nation for managers, technicians, and skilled workers."[40] This change in attitude may have reflected the basic differences between the elite education of President M'Ba at Ecole Montfort, which had emphasized classical parochial teachings, versus the state education of president Bongo at the Brazzaville Technical Academy, which emphasized contemporary skills acquisition. On a more practical level it probably also came about to redress the low level of Gabonese participation in the industrial workforce. According to figures provided in the Second Plan, food industries in Gabon were only 49.5% African, mining industries were only 30.8% African, mechanical and electrical industries were only 40.5% African, textile industries were only 39.7% African, and the other diverse industries were calculated to be only 42.8% African (p. 6). To redress these imbalances the Second Plan intended to allocate 301.9 million CFA francs toward the creation of l'Ecole Normale Régionale d'Enseignement Technique Experimental, and a total of 471.9 million CFA francs on technical education more generally defined (p.118).

The Second Plan introduced many new industrial projects to the government's developmental program. Some of these projects were outgrowths of the country's refining industry, such as the proposed ammonia factory at Port-Gentil, and the Société Gabonaise de Plastique (SOGO-PLAST) factory. Others, such as the carbonated beverage plant in Libreville under the Société pour l'Expansion des Boissons Hygiéniques au Gabon (SEBOGA), or the cigarette factory under the Société des Cigarettes Gabonaises (SOG-ICAL) were designed to meet growing domestic markets. A fishing industry was seriously considered, and based on a protocol accord that was signed between the government of Gabon and the Kawakami International Trading Company of Japan, resulted in the creation of the Pêcheries Industrielles Gabonaises (PECIG) in 1969 with a five-year plan for the construction of flotillas and refregiration entre-pots at the Owendo industrial zone (pp. 51-65).

But, viewing the recapitulative tables on industrial investment, it soon becomes clear that the bulk of government finance in the industries of transformation would go to the cellulose factory (79%). Other mainstay projects such as an extension to the petroleum refinery at Port-Gentil were also earmarked for a disproportionate amount of the total public-private funds. Out of 30.3 billion CFA francs allocated for industrial investment in the Second Plan, 27.8 billion CFA francs—or 91%—was to be spent on the cellulose factory and the refinery's new extension (p. 65).

An effort is made in the Third Plan (1976-1980) to explain why the state chose to distort its industrial investments in favor of the industries of "valorization." (A "valorization" industry can be defined as one that creates "value added" to natural resources.) According to the basic logic of the planners, these industries were simply better business opportunities. Repeating the credo that "industrial development is not an end in itself" the planners set forth in the Third Plan the real objective for which industrialization was merely a means: The primary objective of industrial development was the "augmentation of national revenues."[41]

According to the planners, industries of import-substitution were incapable of providing the increased national revenues of the "valorization" industries which produced for export. One of the main reasons that import-substitution industries could not generate big revenues was because "the domestic market is very limited" (p.133) Another reason was that, given this limited market, "the greater part of the profitable activities have already been set up." As for alternative export-oriented industries, there were additional problems associated with high factor incomes: "The factors of production (manpower, construction costs, energy) and of commercialization (ocean cargo rates) are much more expensive for Gabon" than for other African countries (p. 133).

The secondary objective of industrial development, according to the Third Plan, was to reach a better distribution of revenues throughout the territories. "Valorization" industries that transformed agricultural products (e.g., rice, palm oil, and sugar factories) and wood industries (e.g., cellulose) were seen as better—by virtue of their locations, which were rural and decentralized throughout the territories—than local urban factories which concentrated jobs and incomes in the major cities such as Libreville and Port-Gentil. In this way rural development plans and industrial development plans overlapped. (Keep in mind that the stated objectives of the Third Plan emphasized the need to reduce economic dualism between town and country.)

Notwithstanding the above justifications, what the funding distortion really accomplished was *a perpetuation of economic dependency on primary production industries.* Furthermore the lack of import substitution industries literally doomed Gabon to a future of economic dependence on foreign imports. Moreover, the inclusion of petroleum refining and wood processing as industries—which technically they are—belied the fact that little of what is usually understood as industrialization (i.e., transformation) was not occuring. Adding a saw mill to a forestry enclave was just not the same thing as modern industrialization.

The situation of several of the new industries was not so good. Looking at the fishing industry, the Third Plan found that the production of fish had not stopped falling since 1973. In that year production stood at 4,400 tons; a year later it stood at 3,300 tons, a year later it stood at 1,500 tons (p. 154). The insufficient supply of ice by SOGAPECHE, the difficulty in recruiting manpower, the relatively low price of fish on the markets, and other problems were listed for this flagging industrial venture. Once an industry had been identified by the planners as a profitable venture, the response to its failure was to increase investment and thereby solve its problems. The Third Plan, in this light, was eager to save the fishing industry by allocating around 2.25 billion CFA francs in investments for port facilities and an expanded tuna flotilla.[42]

The forestry-related industries were another group of ventures into which the government planners had decided to channel investments by reason of their profitability. And while it is true that the market for *okoumé* had remained constant over the years, demand for other tropical woods was not as great. "Unfortunately the woods with the highest actual demand on the export market are not those most abundantly represented in the Gabonese forests" (p.161) Therefore the problem for Gabonese forestry was how to encourage growth in its wood industries (sawmills, plywood factories, paper mills, etc.) without over-harvesting the *okoumé*. The Third Plan attempted to solve this problem by allocating around 2.5 billion CFA francs for investments in new sawmills that could handle other woods, in addition to *okoumé*, which could provide inputs for the "valorization" industries at home (p. 162). But at the same time it continued to channel investments into the *okoumé*-based plywood industry: 450 million CFA francs for the *placage deroulé* and 2.5 billion CFA francs for *contreplaqué* itself (pp. 163, 165). No new funds were allocated for expansion of the SOGACEL paper mill, because of difficulties in collecting adequate supplies of wood pulp. The government had modest expectations for its primary transformation industries of wood, which must be distinguished from the forestry sector more generally (i.e., raw *okoumé* exports): Sawmills were expected to increase their value-added production from 1.38 million CFA francs (1973) to 4.90 million CFA francs (1980). Wood veneer factories were expected to increase their value-added production from 1.20 million CFA francs (1973) to 3.06 million CFA francs (1980). Plywood factories were supposed to increase their value-added production from 3.84 million CFA francs (1973) to 8.12 million CFA francs (1980). Secondary transformation (e.g., furniture, train-track ties, etc.) were to increase from .18 million CFA francs (1974) to 1.56 million CFA francs (1980). Paper pulp from nothing (0) to 15.71 during that same time period. Altogether the wood transformation industries were supposed to generate 36.84 million CFA francs by 1980 with a rate of growth of 26.5% during 1974-1980 (p. 167).

The hidden cost of protectionist tariffs gave the food industry in Gabon an appearance of relative prosperity. Collectively those factories and mills that manufactured food and beverages had increased their value-added production 8.4 billion CFA francs in 1974, and had experienced high rates of growth over the previous five years. For example the flour mills had grown at an annual rate of 9% during that time period, the carbonated beverages factories by 11% per year, and the bakeries by around 20% annually (p. 169). But as the planners acknowledge, most of this growth resulted from the economic boom and was "tied to the urbanization and very strong progress of national salaried employees and expatriates" (p. 169).

The Third Plan was really a product of the 1973-74 oil boom. Its efforts at industrial development, both creative and expansive, had been premised on a newfound oil income resulting from the skyrocketing price of petrol on the world markets. The government could, when it chose to do so, channel investments into the creation of food industries because it had the rent to invest. The government could also afford to bolster these import substitution industries with protective tariffs because it had the rent to subsidize domestic industries. So long as the petroleum rents were good, and the government budgets were fat, there was plenty of money to subsidize industry. But even the planners knew that the past growth trends in the food industries were only a temporary phenomenon, whose future was severely limited "in light of the foreseeable reduction in the present economic boom" (p. 169). Food processing industries collectively were forecast a modest 9.2% rate of growth over the course of the Third Plan (p. 171).

The production of beer, for example, had already outmatched the local demand. Beer was fabricated locally by three breweries in Libreville, with a fourth brewery under construction at Owendo by the time of the Third Plan's publication. Consumption in Gabon was expected to reach 55,000 hectolitres by 1980. Meanwhile production capacity of these four breweries was expected to reach 800,000 hectolitres by that same year. Consequently the Third Plan did not foresee any new develops in the brewing industry, although a study was under way on the possibility of fabricating carbonated water from a source at Lecoui. Even if new investments could be justified in this industry, they would not create a substantial number of jobs, for the total number of salaried employees in brewing was 100 (p. 170). Growth in brewing had reached its maximum capacity.

In what was perhaps the most visibly modern industrial branch of the country—the chemicals industry—which included petroleum refining and a series of peripheral manufacturers, government planners had forecast a 19% rate of growth over 1974–1980 (p. 175). This industry had sprouted up around the refinery at Port-Gentil and, with the exception of petroleum products, had manufactured goods for the domestic market. Its total production had increased to 18 billion CFA francs in 1974, a figure that the Third Plan intended to triple by 1980 through major extensions of existing facilities. An ammonia fabrication plant was under construction at Port-Gentil, for example, which would use naphta output from the oil refinery to produce around 60,000 tons/year. It was built with investments on the order of 6.5 billion CFA francs, and was expected to produce around 3.7 billion CFA francs a year through ammonia exports (p. 173). The consumption of paint had increased during the construction boom that accompanied the first oil shock, and so the paint factory at Port-Gentil which produced around 900 tons in 1974 was destined for a second unit with a production capacity of 1.500 tons, so that the "national production (of paint) could progressively substitute imports and could attain around 2,600 tons by 1980" (p. 173). Paints, it should be mentioned, are another value-added byproduct of a petrochemicals industry. Government plans also existed for the construction of a lubricant factory and a chlorine unit, with investments of 700 million CFA francs (not including equipment necessary for the fabrication of drums). Also, the Port-Gentil soap factory was due for a doubling of its production capacity by 1980 (p. 173).

Once again, however, we find that the government planners intended to distort their investments in favor of "valorization" industries over "transformation" ones. The single biggest contributor to the chemical branch of the industrial sector was the Société Gabonaise de Raffinage (which despite the "Gabonaise" in its name was majority-held by the French). The SOGARA refinery at Port-Gentil had cracked approximately 857,000 tons of refined petroleum products in 1974 (i.e., benzene, butane, kerosene, bitumen, naphtha, and fuel oil) worth around 16.95 billion CFA francs—or, in other words, around 93% of the total value of all chemical production in that year alone. Government planners intended on expanding production at the SOGARA refinery to 1,900,000 tons, with an estimated value of 42 billion CFA francs by 1980. This was to be accomplished with investments on the order of 14.4 billion CFA francs, which was more money than was to be invested in all the other chemical units combined (pp. 173-175).

Refinement of petroleum products for export was analogous to sawing logs of *okoumé*. It "valorized" the basic commodity but did not really "transform" it the way, say, a furniture factory or a plywood factory would. Also, because the refined products were destined for export (principally to other UDEAC member states), the refinery resembled other "valorization" industries, producing for foreign consumers rather than for input into local industries. "These units are too small to procure a good profitability" wrote the planners, who announced that no future expansions of the refinery were in line, "and a unit of optimal size (20,000,000 tons) would be too important for Gabonese crude production" (pp. 173-175).

Another branch of the industrial transformation sector that had been stimulated by the first oil shock—construction, materials, and quarries—recorded some impressive gains over the previous five years. This branch included the quarries which extracted marble and granite, the fabrication of cement, the making of bricks and prefabricated concrete, and glassmaking. The consumption of cement had increased quite rapidly during the construction boom years: from 55,000 tons (1971), to 66,000 tons (1972), to 91,000 tons (1973), to 114,000 tons (1974), to 180,000 tons (1975): (p. 175). "By itself the fabrication of cement represents 71% of production and 58% of value-added in this branch in 1974" (p. 175).

Uncertainty regarding the duration of the construction boom, in addition to the availability of cheap building materials from abroad, both served to limit the amount of investment that the government was willing to make in this branch of the industrial sector. For example, a study was conducted on the possibility of building a ceramics unit that would be capable of making bricks and concrete products. Yet "[i]ts implantation will not be justified unless it is not necessary to put up a protection unreasonably higher than cement" (p. 176). That is, the government included protective tariffs into its cost analysis of the project. With cheaper ceramic available abroad, why produce them at home? Meanwhile the "valorization" of rocks was earmarked for more investment money, with the Third Plan setting its aim on the building of a new cement factory in Franceville, and an extension of the crushers in Libreville.

Officially the government defined *industries de transformation* as "the [industrial] sector not including electrical energy production and buildings and public works" (p. 179)—which was a fairly wide-meshed net that held almost every kind of industry imaginable. In this discussion we have been making a distinction that the planners were unwilling to make: i.e., the difference between "valorization" and "transformation." There is some recognition of this difference in the summary of industrial development contained in the Third Plan, when the government recalls that "[t]he rapid growth of industry has been founded essentially upon the valorization industries of the forest ... and of petroleum" (p. 180). But for the purposes of planned development, both kinds of activities were lumped together as a single unity under an even more generalizeable category of the *"secteur secondaire"* (including, in addition to industries of transformation, petroleum research and development [not export], construction, electrical energy, and water).

The distorted investments in—and even more distorted contribution of—"valorization" operations contributed to the confusion between modernization and genuine economic development. For, like the export enclaves, "valorization" activities tended to be capital-intensive rather than labor-intensive. Wood and paper industries were expected to provide 17.6 million CFA francs (40.4%) and petrochemicals fabrication was expected to contribute 5.1 billion CFA francs (12%) out of a total of 43.6 billion CFA francs value added by the industries of transformation over the course of 1974–1980 (p. 180). Rather than producing for domestic industries, however, these two branches manufactured for export. And they employed but a few hundred workers.

As mentioned in the previous section, many of these grandiose plans were abandoned when the government adopted the IMF austerity program in 1978. Nevertheless, most of the industrial endeavors were not devastated by the "bitter pill" of austerity like the rural-agricultural programs had been. The food industries, for example, experienced an 8.6% rate of growth between 1978 and 1982, at current prices: from 2.3 billion CFA francs (1978) to 4.6 billion CFA francs (1982). Industrial chemicals, not including petroleum refinement grew at a rate of 18.4% during that same time period: from 743 million CFA francs to 2.1 billion CFA francs between 1978 and 1982. Construction materials grew 14.8% from 1.9 billion CFA francs in 1978 to 4.8 billion CFA francs in 1982. In fact the only industrial sector to experience a decline was petroleum refining, which fell from 8.1 billion CFA francs to 6.1 billion CFA francs over those same five years.[43]

The decline in petroleum refining can be explained by the presence of a new refinery at Douala, Cameroun. Almost all the petroleum products fabricated by the SOGARA refinery at Port-Gentil were produced for UDEAC member states. When a second SOGARA refinery was built, it cut into the market share for petroleum products previously enjoyed by the Port-Gentil facility.

One reason that the industrial sector had not been hit as hard as the rural-agricultural sector by the austerity plan was because the private sector, rather than the public sector, had provided the majority of its investments. *Reducing public expenditures, therefore, did not affect the industries of "valorization" nor those of genuine "transformation" as much as other sectors.*

What is interesting to observe is how the branch of buildings and public works gradually evolved from being a related part of the *secteur secondaire* in the Third Plan to an integral part of *the secteur industriel* in the Intermediate Plan (1980–1982). Building and public works were identified in the Intermediate Plan as the number-one activity of industrial development, providing the "preponderant part

(44%) of the industrial sector" (p. 159). This shift in categories made possible the claim that the industrial sector in Gabon was experiencing growth. By factoring-in public and private expenses for buildings and public works, the Intermediate Plan simultaneously watered down the meaning of industrialization. It became possible to call a construction site an industrial project, and in the way, a construction boom an industrial boom. When the activity of building public works is defined as "industrialization," allocation can easily become confused with "modernization."

The case of the 1977 Organization of African Unity convened in Libreville is instructive on this point. For many observers the lavish expenditures on construction projects in preparation for the OAU summit was the triggering mechanism that conditioned the country's fiscal crisis and concomitant austerity plan. By some estimates president Bongo committed as much as half of the country's annual budget for that year on building hotels and convention centers and other public works designed to herald Gabon's grandeur and modernity. Many of these projects were never finished. Skyscrapers on beachfront property in Libreville give the illusion of modernity, like a stage set in a dramatic production. The question remains: are they symbols or are they manifestations of the modern?

Here we come to the essence of the discussion. *What is meant by "modern" and by "modernization and development?"* In the discourse of development plans such phrases and terms seem to fall to the level of assumption. Yet this is not a question that must remain enigmatic or, worse, unanswered. For the planners are eager to define their terms, and in so doing, to provide us with a standard by which we can evaluate their success or failure.

The objectives of industrial development, we are told in the Intermediate Plan, revolve around gradually replacing oil and wood exports, replacing raw materials production, with something else. "Development" in the specifically Gabonese sense is *an evolution away from being a commodity exporter to becoming a diversified and productive economy capable of providing for its own basic needs* and augmented financial well-being for all Gabonese citizens:

*Les objectifs et la stratégie du développement indus-
triel reposent sur l'idée maîtresse que le pétrole, qui
assure environ la moitié de la production intérieure
brute et 55% des recettes de l'Etat, est une ressource
temporaire, et que l'industrie doit pouvoir prendre à
terme en partie le relai du pétrole.* (p. 161)

What the planners ultimately mean by development is a
gradual movement toward a non-dependent economic condition, one that liberates the country at last from France
and from its dependence on oil and mineral rentierism. Do
beachfront skyscrapers and choo-choo trains accomplish
this task? Do sawmills and refineries?

Economic recovery, after the IMF austerity plan had
reduced Gabon's indebtedness, permitted the country to
rejuvenate its previous efforts at modernization and development. Visible in its allocations are the state's interpreta-
tions of modernization and development. In its investment
program, the state intended to allocate much more public
revenue to infrastructure than to manufacturing enterprises. Over the course of 1980-1982, Gabon planned to
invest 188,202,000,000 CFA francs in infrastructures such
as roads, ports, airports, telecommunications, and the railroad (see above). *This represents sixty-five times the amount
of public revenue that the state planned to invest in the development of industries and the promotion of small enterprises*
(p. 63). Even when we factor in the estimated 18.5 billion
CFA francs of private investments in Gabon's industrial
manufacturing branch, added to the 2.9 billion CFA francs
of public investments by the state, *the total amount of
money channelled into this kind of modernization and development represented a mere 8% of all developmental
expenses* identified in the Intermediate Plan (pp. 63-65).

The state planned to spend a total of 362.5 billion CFA
francs on development of its economy between 1980 and
1982. Alone, the Transgabonais would consume 118 billion
CFA francs, or around a third of that expenditure. But the
secondary sector employed more salaried workers than
either the primary or tertiary sectors. According to figures
provided by the Intermediate Plan, 45,350 salaried workers
were employed in Gabon's various industries and public

works projects, while only 21,180 were employed in primary production, and 29,800 in tertiary activities (p. 66).

The government's attitude toward industrial manufacturing remained a constant. Even though, in his prefatory remarks to the Fifth Plan, president Bongo stated that its "fundamental objective" was "preparation for the after-oil economy [après-pétrole] based on the diversification of our economy"[44] the content of the plan itself did not suggest that industrial manufacturing was going to ever be a significant part of that diversified future. In fact the Fifth Plan suggests that "directly productive investments are approaching their absorption capacity" (p. 87). Reading the industrial development policy of the Fifth Plan is like listening to a tape recording. The same handicaps, the same problems, the same limitations to industrialization are listed: "The feeble dimension of the domestic market doesn't permit businesses to benefit from economies of scale"; or "the costs of manpower are relatively high"; or "the penury of qualified workers, due to the absence of an industrial tradition and the insufficient professional formation are negative effects on productivity" (p. 169). The same objectives, the same distortions, the same programs are suggested: "The major objective will be to increase the value-added industries"; or "the activities to be promoted in particular are ... the wood industries"; and so on (p. 176).

The point being made here is not that these problems are imaginary, nor that these objectives are illusory, but that *for some reason*, there is a lack of vision when it comes to industrial manufacturing and the production of goods. The explanation for this lack of vision can be understood through an examination of mentalities.

NOTES

1.   Howard Schissel, "Iron Horse to Franceville," *West Africa* (September 6, 1982):2280.
2.   Ministère de l'Economie Nationale, le Commissariat au Plan, *Premier Plan de Développement Economique et Social de la République Gabonaise* (Libreville: 1966), p. 292.
3.   *Deuxième Plan de Développement Economique et Social de la République Gabonaise* (Libreville: Ministère du

Plan, 1971), Préface.
4. Marc Aicardi de Saint-Paul, *Gabon: The Development of a Nation* (New York: Routledge, 1989), p. 66.
5. Jean-Paul Sinsou, "Importance stratégique du Transgabonais dans la politique de développement national" *Géopolitique africaine* (December, 1986):43.
6. *West Africa* (July 29, 1974):944.
7. Ibid.
8. Schissel, p. 2281.
9. "Gabon Government Triples Budget," *West Africa* (October 7, 1974):1225.
10. *West Africa* (June 2, 1975):643.
11. *West Africa* (July 14, 1975):811.
12. *West Africa* (July 28, 1975):796.
13. *West Africa* (July 4, 1977):1389.
14. *West Africa* (January 30, 1978):224.
15. *West Africa* (May 8, 1978):910.
16. *West Africa* (September 6, 1982):2280.
17. *West Africa* (January 15, 1979):104.
18. Michael Reed, "Gabon: a Neo-Colonial Enclave of Enduring French Interests," *Journal of Modern African Studies* 25,2(1987):216.
19. *Plan Intérimaire de Développement Economique et Social de la République Gabonaise*, 1980-1982 (Libreville: Ministère du Plan, du Développement et des Participations, 1979), p. 107.
20. "Matchet's Diary," *West Africa* (January 31, 1983):251.
21. *West Africa* (May 7, 1979):820.
22. *West Africa* (January 11, 1982):128.
23. *Cinquième Plan*, p. 12.
24. Allison Perry, "Coming of the Railway," *West Africa* (February 2, 1987):198.
25. Ibid., p. 199.
26. Reed, p. 317.
27. Perry, p. 197.
28. Perry, pp. 198-199.
29. *Trends in Developing Countries* (New York: World Bank Publications, 1981), p. 210.
30. Chambrier, p. 137.
31. Ibid., p. 335.
32. *West Africa* (August 16, 1976):1167.
33. *West Africa* (May 30, 1977):1025.
34. *Economie et Plan de Développement: République Gabonaise* (Paris: Ministère de la Coopération, 1963), p.

22.
35. *Premier Plan*, op. cit., pp. 288-290.
36. Crawford Young, *Ideology and Development in Africa* (New Haven: Yale University Press, 1982), pp. 241-242.
37. *Deuxième Plan*, op. cit., p. 51.
38. *Premier Plan*, op. cit., p. 305.
39. International Labour Office, *Year Book of Labour Statistics* (Geneva, ILO, 1970) "Industrial Disputes: All Divisions of Economic Activity," p. 784.
40. *Deuxième Plan*, op. cit., p. xiii.
41. Troisième Plan, op. cit., p. 132.
42. Ibid., p. 156.
43. *Plan Intérimaire*, op. cit., p. 60.
44. *Cinquiéme Plan*, op. cit., p. 4.

Chapter 6

The Rentier Mentality in Gabon

INTRODUCTION

F. Scott Fitzgerald once said to his friend Ernest Hemingway that "[t]he rich are different from you and me." To which Hemingway replied: "Yes, they have more money." In a similar vein, we might say that Gabon is different from most of its sub-Saharan neighbors in that it has had more money. But whether that makes it better, or just different, is the question.

The previous chapters have reviewed the economic development plans of Gabon in order to better understand the problems that planners have had to face in achieving their own stated developmental goals. It is important to point out that these are Gabonese development goals, not those imposed from outside. We have judged Gabon by its own standards. Our conclusion has been that, despite more than twenty-five years of public finance, the basic problems that Gabonese planners faced at independence in 1960 remain unchanged, and in some cases have gotten worse.

While we remain deeply sympathetic to the Gabonese objectives of development and diversification, it does appear that an attitude or "*mentality*" exists within the logic of the plans themselves in which the allocation of petroleum revenues has been confused with a short-cut to genuine economic development. This confused mentality seems to suggest that Gabon could use its petroleum, mineral, and forestry rents to purchase the farms, the trains, the roads, and the factories that it lacked in its other underdeveloped sectors.

Gabon has confused allocation with development. Liberated from the need to tax its citizens, Gabon became what Luciani has called an "allocation" (rather than a "production") state. By spending what were effectively *unearned incomes* from its oil the state promoted an *institutionalized largesse* which has brought it twice to the brink of ruin, and has forced it to take the bitter pill of two IMF-sponsored austerity plans.

Like fad diets that promise to shed pounds quickly, these IMF austerity plans temporarily reduced government "waste-lines" ... only to be followed by more allocation binging.

THE MYTH OF THE LAZY NATIVE

There is no reason to immediately dismiss the notion that a rentier economy creates a specific set of attitudes—a rentier *mentality*—which is distinguished from conventional economic behavior in that it, in the words of Hazim Beblawi, "embodies a *break in the work-reward causation.*"[1] But we must be careful to qualify this notion with the priviso that it is a mentality adopted by those individuals who have access to economic rent, the genuine rentiers, rather than some Orientalized misconception of a "lazy native."

Racialist stereotypes of the Gabonese as "lazy" have a long history. One can read them in the letters from the captain of the frigate Masson, commandant of Gabon, to the French Ministry of the Navy and the Colonies, on September 18th, 1882, in which the following remarks were made:

> The Gabonese have become so lazy that if the Pahouins (who are the future because they are energetic workers) did not bring in bananas and

sticks of manioc, the former would have nothing to eat.[2]

Or one can equally recall the muted racism of Dr. Albert Schweitzer, who wrote in his *African Notebook* that his African male nurses shunned good, honest, manual labor: "That a Negro may believe himself to belong to better circles of society is apparent even today when he will not paddle a canoe with others," Dr. Schweitzer tells us.[3]

As S.H. Alatas has explained in his *Myth of the Lazy Native*, such ideas were propagated by the colonizers and often perpetuated by the nationalist bourgeoisie but bore little factual basis. Alienation and withdrawal from forced labor is not the same thing as laziness. Cajoling the colonized peoples by calling them lazy was simply another of the many tools used by the European to cooptate the colonized into participating in the general imperial project.

Notwithstanding the most blatant racialist assumptions, few would deny that something corrosive has happened to the work ethic in Gabon since the oil boom of the 1970s. We suggest that this is related to the rise of the allocation state in Gabon. We must also emphasize that it is a process that started at the top of the social and economic hierarchy and has worked its way down, like acid that is poured on a pyramid, corroding the capstone first, then running down the sides to the base of the structure itself.

The point of discussing the rentier mentality in Gabon is not to provide some kind of totalizing argument about how something in the national culture keeps it underdeveloped. Those familiar with the work of Raphael Patai on the so-called "Bedouin ethos" might recall his attempt to do just that when he categorically annouced that the "[a]version to manual labor, in particular work that involves dirtying one's hands, is another Bedouin attitude that has widely influenced the Arab mind."[4] This quite simply will not do. It is not an adequate explanation for what has plagued economic development plans in Gabon.

Most Gabonese have to work very hard to make a living and survive in their tough economy, as do, for that matter, most Arabs. It is unfair and perhaps even a little bit *conceptually* lazy to assume that a mentality could exist for any large group of people the size of a nation. This is especially true in the case of Gabon, where several hundred ethnicities preclude any kind of meaningful national category.

WHERE IS THE MENTALITY STRONGEST?

Beblawi does provide us with some guideposts when he describes the government of a rentier state as the principal recipient of external rent—a feature which is closely related to the fact that only the few control it. If we are looking for a mentality that is conditioned by economic rent, then we must first locate those individuals and groups with direct access to that rent. In the case of Gabon, where one lives (location) and what one does for a living (occupation) are major factors determining one's access to the rent circuits.

Three regions of the country—Estuaire, Ogooué-Maritime, and Haut-Ogooué—have been estimated to have received around 63% of the state's revenues.[5] That is, people who live in these regions get more rent than those who don't.

These regions have further concentrated public spending in the major cities which they contain: Libreville (the political capital), Port-Gentil (the petroleum capital), and Franceville (the minerals capital). That is, people who live in these cities get more rent than those who don't.

Where one lives, therefore, is a major factor determining access to the rent circuits. Living in Libreville gives one the greatest access. For example, in an article published in *Jeune Afrique* by French journalist Pierre Péan, it was observed that "[t]he money income of an inhabitant of Libreville is about twenty two times that of a traditional agriculturalist."[6] Since the main employer in Libreville is government, government jobs account for most of this income inequality.

Development plans create government jobs. Therefore, development plans have contributed to the income inequality between the town and the countryside. Since independence in 1960, the government bureaucracy in Gabon has swelled at a tremendous rate. So have temporary public worker payrolls. According to Roland Pourtier, the government payroll in Gabon accounts for about a third of all salaried workers. "This phenomenon, which has affected all African states after independence, has taken particularly striking proportions in Gabon."[7] In 1965 there were 3,842 civil servants (*fonctionnaires*) and zero temporary state worker (*main d'oeuvre non permanente de l'Etat*). By 1980 there were 22,173 civil servants and 13,306 temporary state workers. By 1989, Pourtier notes, "[o]ne could estimate that today around 45,000 salaried workers (of which more than 1,500 are expatriates) depend on the state, without counting the para-public sector" (ibid). By 1995 the proportion of state-funded workers out of total employed Gabonese was just as great.

These numbers are relatively high when one considers the small size of the population and territory of Gabon. And not only are the numbers of employees high, but the salaries that they are paid appear, at least to some observers, to be inflated: "With a population of only 1.2 million," observes IMF economist Prosper Youm, "Gabon's civil servants are among the highest paid in Africa, earning the equivalent of those in Senegal which has 7.2 million people."[8] In addition to high salaries, these public employees received many "perks" associated with a good government job: e.g. "[f]ringe benefits like free housing and cars for civil servants," which were only cancelled during the second austerity plan imposed by the IMF in 1987.[9]

Since government has been the primary recipient of Gabon's petroleum rents, it is not surprising that Libreville has enjoyed lavish expenditures unknown to the poor rural villages. "Hardened oilmen become euphoric at the prospect of a posting to Libreville" writes Diana Hubbard of the *Financial Times*, who describes the capital city as a "miniature Nice" where "a woman can walk down the main street without clutching her handbag and the French restaurants would make the Michelin guide."[10] Libreville has been ranked as the third most expensive city in the world for tourists and visitors.

But while the elites and foreigners dance the night away at Fizz & Jazz, le Topkapi, and le Night Fever, the economy of Gabon has been going down the drain. "Gabon's wealthy elite flaunt ostentatious villas and flashy cars," reports *Africa News* "but the vast majority of Gabonese are poor."[11]

This is an illustration of the rentier mentality: a frame of mind which is inappropriate to the circumstances. Michael Reed has observed that the Gabonese view themselves as affluent Africans. And rightfully so, considering their high per-capita national income. "Their occupational dream," writes Reed "often caricatured by the Gabonese themselves, is to wear a suit and tie, carry a briefcase, and work in an air-conditioned city office."[12] This mentality can be seen in the increased share of the labor force that works in the service sector, which has grown from 9.5% in 1965 to 13.8% in 1980.[13]

The rentier mentality creates a break in the work-reward linkage, *making wealth seem like an isolated fact, rather than the result of a long hard process of sacrifice and effort.* One good example of this mentality in action is the way in which wealth is acquired through special situations in a rentier economy created by the sudden influx of external rent. These special situations arise because money which did not previously exist within the economy suddenly does—and it is all concentrated at one single outlet: the state. Without this high concentration of wealth it would not have been possible for *2% of the population to acquire 80% of the national riches.*[14]

As we have stated, the fact that government is the primary recipient of the external rent is closely related to the fact that only the few control it.

What are these "special situations?" One kind—government employment—has already been identified. Without the gargantuan amounts of oil revenue pumped into the government budgets, most Gabonese civil servants presently on the state payroll would be unemployed, and those who were not would be earning far less than they presently do. Considering this almost self-evident fact, it might be interesting to note that many cabinet portfolios and other high-ranking government positions have been given to personal friends and relatives of President Bongo. Some of these individuals are identified in the work published in 1983 under the title, *Les Elites gabonaises* (which can be found in the Library of Congress, for those interested in pursuing this matter further).

Institutionalized corruption is illustrated in the following account provided by Pierre Péan in his *Affaires Africaines*. It is exemplary of how a "special situation" supplies its holder with material wealth. According to an anonymously published report by a group calling itself the "Patriots for the National Health," dated December 5th, 1979, president Bongo is charged with using his presidential power to acquire for himself shares in almost all the major corporations created during his tenure in office.

Whenever a firm wanted to get licensed, according to these sources, that firm had to make so-called "contrbutions" to the president's personal bank accounts. In this way president Bongo and his helpers acquired a shareholding in the following corporations: Sciages Industries de la LOWE, Savonnerie du Gabon, Chimie du Gabon, Grands Garages Gabonais, SOGEMAT, SOMATEM, SOGI, SOGADI, SOGASIG, SOGADIEX, SOCOBA, EDTPL, SODER Gabon, SOSEGAB, SOTRAGO, UDEC, SARIGA, Union Gabonaise de Travaux, Gabon Engineering, SODEC, SETEG, SOGANET, ADG, Air Service, Air Inter-Gabon, Air Affaires Gabon, SOMARGA, SOMICOA, Banque du Gabon et du Luxembourg, Pays-Bas Gabon, Crédit Immobilier du Gabon, SOGAFINEX, Assurances Générales Gabonaises, Omnium Gabonais, Société Gabonaise de Services, SOCOBA, SOACA, and many other companies that collectively comprise the Gabonese industrial manufacturing and commercial base. The total estimated value of all the companies possessed by the president was somewhere on the order of 8 billion French francs, which, when combined with his personal real estate holdings, makes him one of the richest men on the continent—by virtue of his "special situation" rather than hard, honest work.[15]

It is being argued in this section that such a man will not develop realistic ideas concerning economic development. The idea of a rentier mentality suggests that president Bongo would be prone to confusing allocation with real economic development because he personally did not have to deal with the hard work of earning a personal fortune but rather was able to acquire one through the fortuitous circumstances of an oil boom and the disreputable practices of a dictatorial regime.

This supports the World Bank's harsh indictment of corrupt government practices in Gabon. "During the period of oil prosperity," the Bank writes in its report on Gabon, "a policy of state largesse became institutionalized." This state largesse, we are told, was "due to the massive political interference in management, high operating costs, and gross overstaffing" of the 65 parastatal companies that were "the main recipients of oil rent."[16]

An entirely different portrait of Gabon is being painted here than that provided by the development plans. In the development plans we are informed that the government is going to rely on "private initiative" for industrialization. In the development plans we are not informed that the most powerful members of the Gabonese government are also principal shareholders in these so-called "private" industries. Is it any wonder that the plans constantly repeat the same complaints about the same difficulties in achieving profitable import-substitution industries? When these industries must from the outset pay massive extortion money to corrupt government officials even before they produce their first unit?

A report on the Bongo regime in the American press offers a view of the regime's worst attributes during what is decidedly its best economic times, thereby striking the contrast between mere riches and genuine development:

> President Omar Bongo of Gabon is enjoying that rarest of blessings—a second chance. Four years ago he squandered 2 billion dollars on the most expensive bash in the history of African summitry, built 50 villas, a push-button palace, four-lane highways and one of the world's most modern theatres; he also imported a fleet of Rolls Royce limousines and armor-plated Cadillacs. The meeting was impressive, but it almost bankrupt Gabon ... The cut in public funds has left the capital littered with the skeletons of Bongo's former pretensions. The shell of the unfinished Sheraton Hotel looms ominously over the palmfringed coastline. And one four-lane highway now ends abruptly in the bush....

A showman who adores beautiful women and Napoleonic-style cloaks; he rules from a seaside marble fortress which was built, according to some insiders, at the urging of his mother-in-law, Mme Véronique, who complains that spirits from an ancient tribal burial ground haunt the other palace on the outskirts of the town. ... Bongo has stashed away 500 million dollars for himself ... owns two airliners equipped with king-sized beds, a 2.2 million dollar mansion in Hollywood and vast holdings in Gabon itself—and many of his ministers have also enriched themselves.[17]

The last item on this list of grievances is closely related to the personal kinship systems in place wherein "obligation to family and clan determines that the newly arrived relative receive consideration for a position in the office or factory regardless of official merit and performance."[18]

THE TRICKLE DOWN EFFECT

The president is not alone in violating the public trust. His behavior sets the standard for the entire government and has provided a model for such activities as the corrupt "policy of granting vast tracts of virgin forest land to well-connected Gabonese nationals who subsequently farm them out to foreign interests."[19] Fortunes are made in real estate in this fashion, but, in the process, capital is not invested in productive activities and as a result the economy stagnates. Using their "special situations" these well-connected Gabonese elites demonstrate yet another break in the linkage between work and reward.

Anthropologist Michael Reed has observed that the Gabonese themselves are "remarkably uninvolved in any indigenous private commercial sector."[20] This observation is not new. Long before it achieved independence, the country had been dominated by the French business interests, who still continue to control the major commercial enterprises. Few Gabonese were involved in the private commercial sector then. But what Reed finds objectionable is how ever the smaller, or *petit commerce*, is run by expatriates from Cameroun, Nigeria, Senegal, Guinea-Conakry, and other parts of Africa. Foreign merchants have filled a vacuum that ought to have been filled with hard-working Gabonese. In the last two development plans the government tried to promote small and medium-sized enterprises through a state venture capital branch that it called Promogabon. But like other efforts to spend petroleum money in order to purchase development, it has done little to replace expatriate merchants from domestic markets.

The state communications operation, Radio Libreville, was reported to have expressed deep concern about the degree of absenteeism afflicting the Gabonese administration. In a broadcast recorded in October of 1973, "out of 100 officials in a department only 10 were found at work and those in minor positions." According to this report, those individuals in charge were absent and hundreds of important items of work on which the development of Gabon depended "waited for attention."[21]

More recently, when the oil pipeline was being constructed from Rabi field to the Cap-Lopez terminal, the pipeline contractor had problems finding experienced or willing employees to work on the site. There were so few pipeline workers that Elf and Shell subcontractors were competing with one another for the handful that were willing to do this dirty and arduous work.[22] Keep in mind that this was for a project which, by itself, was going to save the national economy. Earlier administrations dealt with similar labor shortages by using prisoners to do the work in the absence of viable alternatives.[23]

When the economy has suffered brief periods of economic contraction, however, a harsh xenophobia emerges in Gabon. Immigrants are blamed for the lack of jobs. In May of 1979, for example, the Lebanese merchant community of around 600 was attacked by vandals and suffered violent looting after a speech by president Bongo blamed unemployment on foreigners. The same sequence of events repeated itself with other expatriate merchants in February of 1985.

Consumer tastes are another reflection of the rentier mentality in Gabon, especially the new tastes for imported foodstuffs that have turned the country into one that is almost completely dependent on foreign imports to meet its basic food requirements. Before the oil boom in the 1970s, Gabon was totally self-sufficient, for example, in its production of bananas. Bananas were a part of the staple diet, and grew wild in the forests of the country. But by as early as 1981 the country had become almost completely dependent on imported bananas from Cameroun: "[W]ith so many farmboys decamped to the cities, they go unpicked."[24]

When these youths arrived in Libreville, however, they were rudely awakened to the fact that a "cheese sandwich and a beer can set you back as much as 10 pounds."[25] The mere fact that a cheese sandwich and a beer were being substituted for bananas epitomizes the changing tastes of an increasingly urbanized population. During the peak years of the oil boom in the early 1980s, Gabon imported 96% of its food from abroad.[26] In US-dollar equivalents the total food import bill averaged around 170 to 180 million U.S. dollars during the remainder of that decade.[27]

One of the most obnoxious symbols of conspicuous consumption of imported foodstuffs was Gabon's seemingly unquenchable thirst for champagne. According to figures put out by the Comité Interprofessionnel du Vin de Champagne, Gabon was the world's leading per capita consumer in 1984, "with one bottle for every three inhabitants."[28] At one wedding held for a prominent political figure's daughter, 8,000 special magnums were specially imported and consumed (ibid).

The decaying work ethos has been observed not just by foreigners. President Bongo himself has repeatedly appealed to his fellow countrymen to be "disciplined and hard working" if Gabon is ever going to pull out of its economic tailspin.[29] Yet his personal example has not been an inspiring performance. And the economy of Gabon itself may not provide the kind of opportunities it once did, when the president could proudly boast that "Gabon is one of the rare countries in the world where everybody who wants to work can do so."[30]

The booming days are over now, and with them the mentality that they created. Today's average Gabonese is deeply concerned with finding and keeping work. In a Marcomer Gallup Institute opinion poll conducted in March of 1988, nearly half the population answered that it was worried about the country's 13% unemployment rate and high food prices.[31]

The problem facing Gabon's workforce may no longer be their supposed unproductivity, but rather, that there is not enough decent work to earn a living wage. For those who have been attending to the task of basic economic development in Gabon, the *après-pétrole* may have come sooner than they expected. Elites previously preoccupied with consuming the national wealth must now find ways of creating it. And if the ideas of the ruling class have become the dominant ideas of the whole society, whence and whither this new national income?

NOTES

1.　[Italics added by the author.] For a discussion of the rentier mentality, see chapter on "Rentier Theory."
2.　*National Colonial Archives*, Gabon-Congo I, dossier III, 21b, cf. Weinstein, *Nation-Building on the Ogooué*, p. 38.
3.　Albert Schweitzer, "On the Tracks of Trader Horn," *African Notebooks* (Indianapolis: Indiana University Press, 1939), p. 26.
4.　Raphael Patai, *The Arab Mind* (1973), p. 113.
5.　Paul Michaud, "Gabon, dynamic television," *West Africa* 3590 (1986):1308.
6.　Pierre Péan, "Le Gabon à l'heure du pétrole," *Jeune Afrique* (June 27, 1975):22
7.　Roland Pourtier, *Le Gabon* (Paris: Harmattan, 1989), p. 205.
8.　"Gabon: Oil Revenues to Spur Diversification" *Africa News* (December 9, 1991).
9.　Lucien Minko, "Gabonese Leader Urges to Break with Oil Dependency," *Reuters* (August 17, 1987).
10.　Diana Hubbard, "Gabon's Oil Refuses to Run Out," *Financial Times* (August 24, 1989):26
11.　*Africa News* (December 9, 1991).
12.　Reed, *Gabon: a Neo-Colonial Enclave*, p. 302.
13.　World Bank, *Trends in Developing Countries*, p. 214.
14.　Francis Kpatindé, "Le Gabon entre Bongo et les Bûcherons," *Jeune Afrique* (November 7-13, 1990), p. 29.

15. Pierre Péan, *Affaires Africaines*, pp. 102-105.
16. World Bank, p. 210.
17. Ray Wilkinson, *Newsweek* (January 18, 1982).
18. James Barnes, *Gabon: Beyond the Colonial Legacy* (Boulder: Westview Press, 1992), p. 115.
19. Howard Schissel, "Gabon Clears Away the Dead Wood," *Financial Times* (September 1, 1989):36.
20. Reed, p. 285.
21. *West Africa* (October 29, 1973):1542.
22. *Pipe Line Industry* (January, 1990):23.
23. *West Africa*, (April 7, 1975):407.
24. *The Economist*, (December 5, 1981):89.
25. Hubbard, op. cit.
26. *The Economist*, op. cit.
27. World Bank, op. cit.
28. *West Africa* (June 3, 1985):1113.
29. "Gabon President Seeks Solidarity in Economic Crisis" Reuters (August 17, 1988).
30. *West Africa* (September 17, 1973):1311.
31. Lucien Minko, "A Changed Gabon Fetes 28th Anniversary of Independence" *Reuters* (August 15, 1988).

Chapter 7

African Rentiers: Nigeria, Angola, Cameroun, and Congo

CONCLUSION: GABON = RENTIER STATE

From the foregoing discussion it is clear that Gabon is a country suffering from considerable social and political duress—a great deal of which can be directly or indirectly related to its primary patterns of dependency. Rather than trying to disentangle these two dependencies, we have integrated them and have concluded that this form of *double*-dependency has made Gabon what it is: The Gabonese polity is deeply dependent on historical institutional linkages with France, linkages that have perpetuated a dominance-dependence relationship and inhibited Gabon's potential to evolve into a free and independent nation state. Simultaneously the Gabonese economy is dependent on external petroleum rents—rents that have both distorted Gabon's patterns of economic development and reduced Gabon's economy to the unfortunate status of mere mining enclave. Gabon would not be the country that it is today were it not for France; nor would it be what it is but for oil.

As for the impact of oil on the nature of the state, the evidence of the discussion indicates that Gabon has in fact met the four characteristics by which Beblawi classifies a rentier state:

1. The Gabonese economy (of which the state is a subset) is one in which "rent situations predominate." It is difficult to delineate with great precision or accuracy the conceptual boundaries of a rentier economy. It is difficult to provide an exact quantitative threshold beyond which an economy is rentier and before which it is not. Nevertheless the evidence of this study leads to the firm conclusion that the case of Gabon falls somewhere within the category of the rentier economy. Petroleum rents have provided, provide, and will continue to provide the lion's share of both the Gabonese national income and the Gabonese public finance regimes throughout the remainder of the twentieth century and beyond. That is to say, not only does petroleum account for the bulk of Gabon's gross export, gross domestic, and gross national product, but oil rents also provide around nine-tenths of the Gabonese operational budgets and the totality of the financing for Gabonese social and economic development plans.

2. The origins of Gabonese oil rents are "external to the economy" and not domestically generated. That is to say, the predominant source of Gabonese wealth is not derived from the productive activities of the Gabonese people themselves (i.e., from the efforts of any domestic Gabonese capitalist class), but instead derive from overseas exogenous origins, coming principally from France, but also from Britain, the Netherlands, and the United States. Gabon is therefore a country dependent upon and vulnerable to the fluctuations of the world markets and to the variable demands of foreign consumers—a country where what

happens inside its borders is determined by what happens outside its borders; a country that lacks control over its own survival; an object acted upon by the world system rather than a subject acting within it.

3. Only the few are involved in the generation of rents in Gabon, while the majority are involved in its distribution and consumption through public finance, operations budgets, and development plans. The monopolization of Gabonese oil enclaves onshore and offshore by the Elf-Gabon, Shell-Gabon, Amoco, Tenneco, and a handful of foreign-operated firms has effectively isolated the single most important industry in the country from the vast majority of the population. The pattern has been one of enclave industrialization, demanding little of the domestic Gabonese output in the petroleum developmental process, and supplying little of the petroleum for input into the activities of the domestic production and manufacture. Instead of rapidly developing into a modern industrialized economy, after three decades, the Gabonese "modern" sector—if we can call it that—has merely changed outward appearances, from a sleepy timber enclave to a restless oil enclave. Frenchmen run the key operations. Few Gabonese are directly employed in these oil enclaves, which stand like islands of gleaming prosperity and high-technology amidst a sea of poverty and underdevelopment.

4. Finally, the government of Gabon is the principal recipient of the external rent in the economy. This is closely related to the fact that few are involved in the generation of rents. The Gabonese petroleum sector, run by a handful of foreign engineers, diplomats, ex-soldiers, and mercenaries, has bought and paid for the development of a Gabonese rentier class. This class runs the bogus apparatus of government for itself and for Elf-Gabon, and has consistently resisted efforts at democratization or any substantial change to the status quo. High political office has become the road to riches in Gabon, and the struggle for political power has been turned into a struggle for economic survival. Big oil money has spoiled an already elite political class of *évolués* who had been living in a genteel poverty in Libreville, turning them into a *nouveaux riches* rentier class, living a life of conspicuous consumption and dangerous

leisure. Big oil money has also transformed the function of the Libreville governmental apparatus from the caretaker role it inherited from its predecessor French colonial administration into a veritable "sugar-daddy" allocation state.

The evidence also suggests that Gabon has become what Luciani called an "allocation" state, one in which government spending has been confused with genuine economic development. In the years to come, it shall be seen whether or not president Bongo's Transgabonais railroad was worth the pact he struck with the devil to complete it. But the hundreds of millions of CFA francs that the Gabonese government expended on rural development projects have already been evaluated and are found seriously wanting. And the hundreds of millions of CFA francs that the Gabonese government has invested in real estate and manufacturing have been liquidated in the IMF austerity plans, which called for disinvestment and privatization in order to receive debt leniency. The end result of billions of CFA francs expenditure is little more than a handful of rusting factories, a choo-choo train, and a massive government debt. Like the profligate son who returned home after spending his entire inheritance, the Gabonese state will soon be forced to return home to a more basic African economic reality.

French imperialist policies are a shared experience for all francophonic black African states. Birthed in a family of colonial France, liberated by a patriarchal Charles de Gaulle, these states remain to one degree or another dependent on the neocolonial motherland. France nourishes her sucklings with aid. But this is a corrupt family system, where dysfunction and abuse stunt maturation under a mantle of massive diplomatic denial. Within the family portrait of France's neoimperial *coopération*, interstate power relations have tended to enmesh Gabonese governmentalities within the logic of French accumulation. From this diplomatic incest what has emerged is an African humunculus. It is more than a matter of France dominating Gabonese petroleum. France also dominates Gabonese trade. France dominates Gabonese monetary policy. France dominates the provision of Gabon's financial aid and technical assistance. France dominates direct foreign investment in Gabon. France dominates Gabonese law. France dominates the Gabonese military and security forces. France dominates the Gabonese education system. France dominates Gabonese political insitutions. France dominates Gabonese economic policy. France dominates Gabon.

What does this domination ultimately mean? How does it affect the nature of the state in Gabon? What is this state? this Gabon? There is a sense that external recognition of sovereignty by the international system may be sustaining hollow regimes essentially lacking internal political coherence. Since nominal independence was granted to Gabon by a grudging France in 1960, our essentially formalist international system has been more than willing to accept the illusion of Gabon: i.e., that it is a real and independent state in a world of real and independent states. Evidence of this projection surrounds us everywhere. Maps of the world show Gabon's territorial borders. The United Nations recognizes the Republic of Gabon and gives it a seat in the General Assembly. Gabon is a member state of OPEC, UDEAC, the OAU, and several other major international and interstate organizations. Gabon prints its own stamps. It has a national anthem. It holds elections and chooses its own government, appoints ambassadors who work in embassies overseas, issues visas to enter and exit its domain—all reinforcing this illusion of statehood. To the degree that our world recognizes Gabon as an independent African state, it would seem that Gabon is one.

NIGERIA, ANGOLA, CAMEROUN, AND THE CONGO

Are there other rentier states in Africa? Mahdavy suggests that to study the structure of rentier states "one may use time series for a single country to show departures from past trends after the external rents have become significant in amount."[1] One might also compare several countries over time to analyze the convergent and divergent trends they exhibit. Further research on rentier states in Africa is warranted by the alarming case of Gabon and can be extended to at least four other countries in the sub-Saharan African region. Not exclusive of other cases, these four countries are: (1) Nigeria, (2) Angola, (3) Cameroun, and (4) Congo (Brazzaville).

Nigeria is Africa's biggest oil producer, pumping on average more than one-and-a-half million barrels of crude per day. From a comparative standpoint, the case of Nigeria, with its population of around one-hundred million, contrasts in magnitude with that of tiny Gabon and its population of less than a million inhabitants. In addition to being Africa's biggest oil producer, Nigeria is also the most populous country on the continent, and a quintessential high absorber of its oil export revenues. "Typically," writes Mahdavy, "the Rentier States are 'small' in size," yet he suggests that Nigeria would soon be "joining the group."[2]

Is the Nigerian oil experience comparable to that of Gabon? Can a country as large and as energetic as Nigeria be sufficiently encapsulated by the rentier model?

Angola is the second largest producer south of the Sahara, producing on average around half a million barrels of crude per day. Unlike the Gabonese and Nigerian governments, which are capitalist in their orientation, the ruling MPLA regime in Angola has been ideologically Afromarxist. Additionally it has chosen to remain outside the OPEC cartel. It is also a regime that has not known peace. From a comparative standpoint, the case of the Angolan state—with its history of armed struggle for independence from Portugal, and then with its long bloody civil war against UNITA rebels—contrasts violently with that of quiet Gabon and its quarter century of stable and uninterrupted civilian rule.

Is the Angolan experience comparable to that of Gabon? Can a country torn asunder by a state of permanent war be adequately described by the model of the rentier?

Cameroun is the most recent West African oil producing state. With an average output of a little under two-hundred-thousand barrels of crude per day, it has quickly risen to become the fourth largest producer in the region. From a comparative standpoint, the late arrival of Cameroun onto the world oil scene contrasts in duration with that of Gabon, where production dates back to the colonial era. The Camerounians have had the opportunity to learn from the bittersweet experiences of their oil-dependent neighbors, and the Camerounian political leadership has actively sought to avoid the pitfalls of oil-rentierism.

Have the advantages of followership offset the debilitating effects of oil-rent dependency in Cameroun? Can a country that is structurally dependent on economic rent politically alter the effects of rentierism?

Congo-Brazzaville is the fifth largest producer in the region, producing on average around one-hundred-and-fifty-thousand barrels of crude per day. From a comparative standpoint, the case of Congo-Brazzaville shares a great deal with Gabon. It has suffered a commom colonial past (both countries were parts of the French Equatorial African federation) and shares a common neocolonial present including a common currency through the franc zone, common military and economic "cooperation" treaties with France, and a common petroleum corporation—Elf—dominating its oil industry. Despite these commonalities, there are still important differences between these two countries. The government of the People's Republic of Congo-Brazzaville is officially Afromarxist and historically unstable, with a successful revolution in 1963 and the execution, assassination, and house arrest of three of its last four presidents.

Is the rentier model able to sum up the Congolese experience as well as it did the Gabonese?

TRANSNATIONAL CONTROL

One of the key questions has been, *"Who controls the development process in these countries?"* Are people the objects of development under someone else's control or are they subjects of development? One of the most striking commonalities of these four states is the domination and control of their oil sectors by large transnational corporations: Shell-BP, Gulf-Chevron, Elf, and Agip.

The first Nigerian oil exploration began in 1937 under the auspices of the D'Arcy Exploration Company.[3] This firm acted on behalf of two parent companies—Royal-Dutch Shell and Anglo-Iranian Oil Company—assuming the name Shell-D'Arcy Petroleum Development Company of Nigeria, Ltd. in 1952.[4] The British and Dutch owned Shell-D'Arcy on a 50-50 basis, and spent more than $31 million on wildcat exploration in the swampy—and at that time relatively inaccessible—part of the Niger Delta.[5] In 1956 Anglo-Iranian Oil Company changed its name to British Petroleum, and subsequently the Nigerian firm became Shell-BP. In late 1951 the company spudded its first successful well, and five years later annouced that a $3.5 million drilling barge had been ordered from a shipyard in Rotterdam with a landing deck for helicopters, floating craft, and other equipment for tidal swamp operations.[6] By 1964, as a result of a cumulative expenditure of $560 million,[7] *Shell-BP produced 95.9% of total Nigerian crude output.*[8] In 1978 Shell accounted for 57.7% of Nigeria's output.[9] More than a decade later, Shell is still the country's largest producer.[10]

The development of the Angolan oil industry started in 1918 when the Companhia des Petroles de Angola (a Sinclair Oil subsidiary) first began prospecting the colony. The great difference between the Angolan case and the other sub-Saharan rentiers was that the colonial Portuguese state was itself underdeveloped and therefore lacked a national oil company of its own to prospect for its imperial oil. The Belgian firm Petrofina was granted an exploration license in 1952 and subsequently discovered southwestern Africa's first oil field of commercial quantities.[11] But the major discoveries were to occur at Cabinda, an enclave territorially separated from Angola and administered by Portuguese colonial authorities as independent from Luanda. In 1957 the Overseas Ministry in Lisbon issued a decree authorizing the American corporation Gulf Oil an exploration concession to this enclave. In 1966 Gulf found oil there and began producing in 1968 under the auspices of its subsidiary the Cabinda Gulf Oil Corporation. By 1973, *89.6% of all Angolan production came from Gulf's Cabinda concession alone.*[12] In 1984, Gulf Oil was bought out by the Chevron corporation, which via its Cabinda enclave operations remains to this day the single most important producer in Angola, generating around nine-tenths of total crude output.

The Camerounian oil industry was much slower to develop than the other African rentiers, yet followed the same overall pattern and consequently evolved into a French-dominated oil enclave much like Gabon. In 1948 the French government granted an exploratory concession to Société de Recherches et d'Exploitation des Pétroles du Cameroun (SEREP-CA) which like SPAEF was a subsidiary branch of the French state oil apparatus. After ten years of exploratory work, because of the firm's limited success, one of the two rigs operating in Cameroun was moved south to Gabon.[13] Exploration was halted altogether during the early divisive years of political independence, not being resumed until 1961. In 1972 SEREP-CA made its first major discovery offshore at the Rio del Rey concession. Production and export commenced in 1977, *all of it by Elf-SEREP-CA, which remains Cameroun's sole producing firm.*

The development of the Congolese oil industry was at first inseparable from that of Gabon. The French government originally granted an exploration concession to Société des Pétroles de l'Afrique Equatoriale Française (SPAEF) for territory spanning both the Congolese and Gabonese coastal regions. Therefore the same company that developed the first Gabonese oil was, on November 1, 1957, also responsible for the completion of the first Congolese well. By 1960 this French firm was producing commercial amounts of oil at Pointe Indienne field on a 100%-SPAEF-held onshore coastal permit. The company title of SPAEF was changed after political independence to SPAFE and then eventually to Elf-Congo. For the remainder of the decade Elf-Congo was the sole operator in the country, virtually monopolizing Congolese crude oil production—until 1969, when Agip Mineraria, the exploration and production wing of the Italian national oil company ENI, acquired an offshore concession in the Congo. Soon the French and Italians negotiated an agreement by which Elf and Agip exchanged 35% share in one another's Congolese subsidiaries, resulting in the creation of two joint-venture firms: Elf (65%)/ Agip (35%), and Agip (65%)/ Elf (35%). The French thereby kept an interest in all oil operations in the country, while the Italians were able to lessen their financial risk. These two companies continue to monopolize Congolese crude oil production. In the late 1980s *the French, through Elf-associated concessions, produced 95% of the country's total output.*[14]

Becoming Dependent on Oil Rent

Another of the early indicators of rentierism was that Nigerian production was immediately geared for export rather than for domestic consumption. Commercial production in Nigeria began in 1957, and the first 65,000 barrels of crude oil were shipped from the Shell-BP terminal at Port-Harcourt on February 17, 1958, to a Shell terminal in Rotterdam. The second consignment of 105,000 barrels was exported to Britain in March.[15] Although Shell-BP invested in a local refinery at Port-Harcourt to meet local needs, the vast bulk of Nigerian oil would continue to be exported elsewhere for foreign consumers.

One consequence of production-for-export was the infusion of external rents generated by those exports, resulting in an overall structural change in the nature of the economy. "The most central fact of postindependence Nigerian political economy" writes Crawford Young, "has been the total transformation from agricultural exporter to major oil producer."[16] Before political independence in 1960 the main Nigerian exports were palm oil (from the Eastern region), cocoa (from the Western region), and groundnuts (from the Northern region). In 1959 a little under $3 million in state revenues came from the new oil exports.[17] By the early 1970s, the country's increasing oil production provided its military regime with $3 million a day in rent.[18] In 1972 Nigeria had become such a significant exporter that it actually supplied the United States that year with more oil than Saudi Arabia. In the spring of 1975, the Obasanjo government re-arranged the Nigerian tax structure, reducing the price for buy-back oil to 42 cents a barrel while at the same time boosting the tax rate from 60.78% to 85%, and royalty rates from 16.66% to 20%.[19] Then, in the first quarter of 1976, the government increased posted prices across the board to $12.75 a barrel for its 55% participation crude (the highest price in OPEC), decreased allowables, and cut company profits per barrel by more than 36% in the first quarter of 1976.[20] In 1969 crude oil accounted for 41.2% of total exports and for only 15.6% of total government revenues. Ten years later, in 1979, *oil production accounted for 93.4% of total exports and for 89.4% of total government revenues.*[21] Following the pattern of production-for-export, the first Angolan crude oil cargo—30,000 barrels from Petrofina's operations 13 miles south of Luanda—was shipped in September 1956 to the lusophonic colonial metropole of Lisbon, in fulfillment of Portuguese dictator Antonio Salazar's plan for the "methodical exploitation of the natural resources of the colony" and "the progressive integration of all economic activities into the Portuguese economy."[22] However, when Gulf Oil started producing oil in the Cabindan enclave in 1966 it was for export to the Western Hemisphere,[23] with the Portuguese government receiving 55% interest in the operations, plus taxes and royalties.[24] After political independence in 1975, when ownership of Cabinda went to the new government in Luanda, Cabindan oil exports would provide $1.5 million per day in rent to the afromarxist MPLA regime, accounting for about half of the country's foreign exchange.[25] "In practical terms," said president Augostinho Neto, "the government has opted for an accommodationist economic policy, especially *vis-à-vis* the multinational corporations."[26] After Neto's death in 1979 and his succession by Eduardo Dos Santos, the war-torn Angolan state still depended on oil rents for practically all of its revenues: *Hard currency revenues from oil exports in 1984 were $1.8 billion,*[27] *representing more than 90% of all exports and government revenues.* According to the World Bank, in 1989, petroleum exports were valued at $2,657,000,000—more than ten times the value of the country's second biggest export, diamonds, which were valued at a mere $229,000,000.[28]

President Neto appreciated the political consequences of economic dependence on oil: "Can we say we are completely independent while Cabinda Gulf Oil exploits the petroleum of Cabinda? Obviously not."[29]

Cameroun attempted to harness petroleum revenues for agricultural and industrial development. According to the International Monetary Fund, agricultural production had provided 41.3% of Cameroun's gross domestic product in 1964, and by itself employed more than 75% of the population. The major export crops were cocoa, coffee, cotton, bananas, rubber, groundnuts, and palm oil. Among the francophone black African states Cameroun ranked second only to the Ivory Coast as the largest exporter of coffee.[30] Camerounian president Ahmadou Ahidjo recognized the importance of balanced development and instituted a series of projects aimed at improving the agricultural production of the nation. He also attempted to gain state control over this state product. Once oil started flowing in 1977, the state garnered a 60% share of all production, which it began feeding into a state-operated refinery in 1981. The refinery met domestic consumption needs of 30,000 bpd, with net refined product sold to Chad and the Central African Republic.[31] But despite the government's efforts to limit the effect of oil rents, petroleum slowly became the dominant export commodity in the economy. According to the World Bank, for example, coffee exports in 1980 were valued at $326,000,000 and petroleum exports at $398,000,000. By 1990 coffee exports had declined in value to a mere $175,000,000 while the value of petroleum exports increased to an extraordinary $825,000,000.[32] The decline in the value of agricultural exports occured while the share of the labor force represented by agriculture remained over 68%.[33] Augmenting this was the fact that Cameroun also became an important regional producer of industrial products, increasing its manufactured exports from $119,000,000 in 1980 to $210,000,000 in 1990.[34] Cameroun became sub-Saharan Africa's third largest oil producer and exporter, and for now, a country substantially dependent on external rent.

Congolese oil followed the pattern of production-for-export of its neighbours to the north and south, with initial output in 1960 being integrated into a French pipeline system that transported the crude to storage facilities between Pointe-Indienne and Pointe-Noire, where tankers were then loaded off-shore at the mouth of the Rouge River, and thereafter sent overseas to Le Havre, France.[35] The "Three Glorious Days" of 1963 had resulted in the overthrow of the pliant francophile Youlou regime and its replacement by a more nationalistic Afromarxist regime. This was when serious rent-seeking began. The Congolese government attempted to wrest control of its finances and trade from the French. Although keeping itself in the UDEAC and the CFA franc zone, "[i]mport-export trade was taken over, and state marketing monopolies were created."[36] Petroleum distribution networks were taken over.

The official Marxist-Leninist rhetoric in the People's Republic of the Congo stood in sharp contrast, however, to the economic realities of a nation dependent on foreign capitalists and economic rent. *In 1983 oil exports provided 80% of total Congolese export earnings.*[37] And while cocoa exports fell in value from $534,800,000 in 1980 to $94,700,000 in 1989, petroleum exports grew in value from $727,000,000 to $873,000,000 over the course of the 1980s.[38] By 1985, about two-thirds of the government budget came from oil.[39]

OIL AND AUTHORITARIANISM

Rentier theory posits a causal relationship between, on the one hand, the material conditions and social forces of production of a country and, on the other hand, that country's political system: i.e., it is assumed that from a rentier economy emerges a rentier state. Having suggested that these four African countries have in fact experienced an increasing dependence on external rent that is *per se* definitive of a rentier economy, is there evidence warranting the hypothesis that their states have become rentier? Are they inimical to the development of political democracy?

The idea is not to blame the lack of democracy and the presence of authoritarianism in Africa on the mere existence of oil—for after all what states in Africa have truly democratic regimes?—but rather is to show that these states conform to the general pattern of the rentier state.

Two of the most important features of the Nigerian state since political independence do fit the mold of the rentier state: (1) its state has been almost continuously ruled by military officers, and (2) its leaders are almost never replaced by electoral-democratic means but instead are either assassinated by political opponents or overthrown by *coup d'Etat*. For example, whereas Alhaji Abubakar Tafewa Balewa and Benjamin Nnamdi Azikiwé were elected into office, both were unconstitutionally deposed: Balewa being assassinated, and his successor General Johnson Aguyi-Ironsi killed in the coup of 1966. General Yakubu "Jack" Gowon was himself overthrown by the coup of 1975, and General Murtala Muhammad who replaced him was assassinated in 1976. General Olusegun Obasanjo was pressured to step down for civilian elections, but the civilian government of Shehu Usman Aliyu Shagari was overthrown by coup d'Etat in 1983, and even his successor General Muhammadu Buhari was deposed by a bloodless coup by General Ibrahim Babangida. Recently held elections were cancelled by the military government, postponing the arrival of democratic civilian rule still longer in this country.

Similarly, the Angolan state has been ruled by an authoritarian military regime, failing to win total national legitimacy through the vehicle of democratic elections. However, in Angola the monopolization of the institutions of state by the ruling MPLA party, and the anti-regime violence and political instability by the opposition UNITA forces in the south, have both been conditioned as much by outside interference in Angolan domestic political affairs as they have been determined by this country's dependence on external rent. Since Angola gained political independence from Portugal in 1975, the diplomatic desiderata of both the United States and South Africa have been to eliminate the Marxist MPLA regimes of Augostinho Neto and Eduardo Dos Santos. Consequently both the U.S. and the R.S.A. have played major parts in supporting UNITA and Jonas Savimbi's bloody campaign of terror. But America's unquenchable thirst for oil has led it into a curious foreign policy. For through the Gulf Oil Company of Cabinda (Gulf-Chevron), the U.S. has simultaneously been providing this Marxist regime with the hard currency it has needed to wage its campaign against Jonas Savimbi and UNITA forces. The resulting destabilization and civil war has contributed to the perpetuation of permanent wartime mobilization, and consequently, to a military regime in Luanda.

In Cameroun the government has been dominated by civilians, but—in a pattern reminiscent of Gabon—dominated by a single political party. The Camerounian National Union (CNU) was created by president Ahmadou Ahidjo, who, after 22 years in office, suddenly announced his retirement in 1982. The Camerounian constitution specified that the prime minister, Paul Biya, was the rightful successor to the president. When Biya assumed the office of the presidency, Ahidjo retained his position as head of the national CNU party. This resulted in a state of political bicephalism that ended only when Biya announced the discovery of a plot against the administration by Ahidjo and certain northern barons, leading to a trial and death sentence (later pardoned) of the ex-president and a purge of many government officials.

Congo-Brazzaville, finally, fits the anti-democratic pattern of the rentier state. As mentioned above, the puppet regime of Abbé Filbert Youlou was overthrown by a people's revolution rather than a democratic election. In 1964, the Congolese people elected Alphonse Massamba-Debat president, who proclaimed a one party regime under the ostensibly Marxist-Leninist Mouvement National de la Révolution (MNR). Attempts to politically indoctrinate the national army, however, provoked a mutiny by northern soldiers personally loyal to Captain Marien Ngouabi, who in 1968 dismissed Massamba Debat from power and abolished the MNR, declaring a new People's Republic of the Congo with a new Leninist single-party regime under the Parti Congolais du Travail (PCT). Marien Ngouabi was assassinated in 1977 and a new military government under General Yhombi-Opango took power. Massamba-Debat was accused of being involved in the assassination, and was summarily executed by the military government in 1977. Yhombi Opango resigned in 1979 under accusations of corruption, and was placed under house arrest and accused of treason by Colonel Denis Sassou-Nguesso, who after 1979 served as the unelected ruler of the People's Republic of the Congo.

RENTIER CLASS, RENTIER MENTALITY

Another feature of a rentier state is the fact that the few control the inflow of economic rent, posing a temptation for the formation of a rentier class, and for personal aggrandizement as the prerogative of political power. In the case of Gabon we saw evidence of widespread corruption. What kind of evidence is there for a similar cleptocratic mindframe in these other oil rentier states?

There are many examples of corruption in the Nigerian political record, but a few will amply illustrate the presence of some kind of dysfunctional mentality. During the Gowon regime, ten state governors were found guilty of corruption for using their offices to enrich themselves with petroleum rents. In 1979 an internal audit revealed that $5 billion were "missing" from the state oil company, Nigerian National Petroleum Corporation (NNPC).[40] "Under Shagari's NPN" writes Othman, "a class of millionaire hucksters rose to political preeminence at the expense of the poor and the possessors of academic, technocratic, corporate or managerial skills."[41] During the oil glut, expatriate oil men were smuggling oil out of the country, creating shortages for many goods, including much-needed lubricants for oil-pumping equipment, the veritable bloodline of the economy. Sayre Schatz labelled Nigerian allocation policies as "nationalists nurture-capitalism with state capitalist and welfare tendencies"[42] which revealed a penchant for corrupt contracting procedures and a proclivity to spend on projects of what he coined "euphoric capitalism." One of the prime examples of the Nigerian state's incapacity to absorb massive amounts of external rent was the so-called "cement scandal" of 1975, when Nigeria ordered 16 million tons of cement, but lacking port space, left over 260 ships lying offshore laden with their cargo. Profligate spending by this military regime drove the Nigerian budget to an 800-million-naira deficit.[43]

Some of the expenditure was on dubious political projects, such as the creation of a new capital city of Abuja in the middle of the country. Other expenditures were on textbook cases of industrial "white elephants" such as the Bonny LNG project. Bonny started when the Nigerian government convinced/deluded itself that there would be an expanded market for liquefied natural gas (LNG)—particularly in the United States—in the 1980s (it is hard, with hindsight, not to wince) and sent representatives to the United States in order to get firm commitments from U.S. companies for Nigerian gas exports. In 1978 the Nigerian

National Petroleum Corporation (NNPC)—which had replaced the NNOC as the state oil firm—reached an agreement with Shell-BP, AGIP-Phillips, and Elf for the construction of a $5 billion LNG facility at Bonny. Shell and BP were to receive 10% each, Phillips and AGIP 15% between them, and Elf 5% for its participation. The remaining 60% would belong to the NNPC, which would also take between 85% and 100% of the pipeline system needed to bring gas from the fields within a 50-mile radius of Bonny.[44] With oil revenues providing 94% of foreign exchange earnings and 80% of federally collected revenues, the coming of the oil price crash of the 1980s spelled economic disaster for Nigeria. The Bonny LNG project was the first agreement to collapse, when Phillips Petroleum withdrew its 7.5% equity, and resigned its role as technical leader. BP was the next to withdraw, followed by the remaining companies in the consortium.[45]

Civil war expenses drained what riches were available for the creation of a rentier class in Angola. Truly it is an example of the national wealth going up in smoke. Unless one is going to count either Cuban troops or foreign petroleum engineers among the Angolan rentier class, the term seems to lose its meaning in the usual sense. Which raises yet again the question of whether or not a country such as Angola, which meets the economic criteria of an oil-rentier state, can be considered as an example of this category? Unlike the other oil states where a class of nouveaux riches emerged from a special access to the rent circuits, in Angola the money went to purchase arms and military supplies for its bloody civil war. The state itself barely seems to have materialized, far less a ruling rentier class. If anything, who shall receive Angolan oil rents may be a latent simmering point of contention that will only be settled when the more basic question of who shall rule Angola is finally decided. For many the end of the Angolan civil war is expected to lead to a new era of prosperity. But, knowing what has happened in other oil-rentier states, one must ask: Will not settlement of the civil war merely enable the trademark aggrandizement, corruption, and venality of civilian rentierism?

In Cameroun the identifiable elites may or may not correspond well with the rentier class model. It was widely known that political power under the Ahidjo regime "rested with a small clique of Northern Muslim barons close to him."[46] As the state became, with the ascendancy of oil export revenues, the primary source of revenue in the Camerounian economy, this structural transformation produced a veritable rentier class. "During the Ahidjo epoch," writes Jean-François Bayart, "it was a linkage to the state which gave actors their capacity to enrich themselves and to dominate the social field."[47] Unlike Angola or Gabon, there did develop in Cameroun a thriving capitalist bourgeoisie. The Douala Chamber of Commerce, Industry and Mines, representing southern business interests of the country, complained bitterly about the abuses and corruption of the Northern-dominated Yaounda government. In a letter addressed to president Biya, the Douala Chamber of Commerce complained about: the "scandalous favouritism in the regulation of goods and services imports"; the "willful disregard of true business to the benefit of adventurers without licenses"; the "tolerance, indeed an encouragement of organized customs fraud"; and the "generalization of contraband trade by land, sea, and air."[48] But to single out the state for self-aggrandizement is unfair, for, as Bayart observes: "At the end of the day bourgeoise and bureaucrats drive in the same Mercedes, drink the same champagne, smoke the same cigars, and meet up with each other in Airport VIP lounges."[49]

The Congo inherited a massive bureaucratic apparatus from Brazzaville's former role as capital city of the French Equatorial African federation. So expensive was the inherited bureaucratic apparatus that the Youlou government had a continuous overall budget deficit: 24 million CFA francs (1960); 486 million CFA francs (1961); 674 million CFA francs (1962).[50] History initially bred a "southern" political class loyal to the French as much as if not more than to the many competing ethnic groups of the "north." Despite their anti-imperialist rhetoric, the succeeding regimes were dominated by southerners and formed close bonds with the imperialists. "Sassou-Nguesso was especially careful to encourage a strong relationship with

France, the Congo's most important trading partner and supplier of aid."[51] Corruption of this political class was not so hard to imagine. "[I]n 1979, the Congolese radio was accusing Yhombi of having embezzled over U.S. $50 million."[52] Nor is it impossible to imagine the development of a rentier mentality in this country where the majority of the citizens are unemployed youths aspiring to find a good government job or some other special situation in the rent circuits that dominated the economy. But imagining these developments and demonstrating them are two different things.

CONCLUSION: FURTHER QUESTIONS

There do appear to be several common patterns worth investigating in the cases of these four sub-Saharan African oil-exporting states. But there are also other African states which do not produce/export oil yet still depend substantially upon external rent for the greater part of their national incomes. One question that remains to be answered is whether there is something special about the oil-rentier states, or whether other rent-dependent countries are conceptually fit for the model of the rentier state? For example, do the diamond-dependent countries such as Botswana and Namibia share any important characteristics of the rentier state model? So long as they depend on the rents generated by their diamond industries, and those diamonds rents are from external sources, it seems as though they would technically qualify.

But unlike diamonds, which have an artificially created and sustained marketplace that has been historically monopolized and manipulated by the Anglo-American corporation (*vis-à-vis* Harry Oppenheimer's De Beers), petroleum has become a central component of the industrialized world which feeds the mechanized and motorized societies of the core of the capitalist world-system with the energy and raw materials of its dominant mode of production. Furthermore the political power of oil is more often suggested than defined. As the single largest commodity traded in the world (in terms of volume), petroleum is often identified as an independent variable in international relations. Some have argued that oil is "a commodity intimately intertwined with national strategies and global politics and power."[53] Others have isolated oil as the cause of several wars and conflicts and a factor in strategic policy and actions.[54]

We live in a world veritably running on oil. Those states which are integrally linked to its production and export may very well occupy a unique position in our present international system. Or they may be merely the most extreme yet still typical examples of the kind of enclave economic systems that one would expect to find increasingly manifest in the capitalist world-system of the late twentieth century and beyond.

NOTES

1. Hossein Mahdavy, "Patterns and Problems of Economic Development in Rentier States: The Case of Iran" in M.A. Cook, ed., *Studies in the Economic History of the Middle East: From the Rise of Islam to the Present Day* (London: Oxford University Press, 1970), p. 438.
2. Ibid., p. 431.
3. *World Oil* (August 1949).
4. *World Oil* (August 1952).
5. *World Oil* (August 1956).
6. *World Oil* (August 1957).
7. *World Oil* (August 1964).
8. *World Oil* (August 1965).
9. *World Oil* (August 1979).
10. *World Oil* (August 1990).
11. *World Oil* (August 1953).
12. *World Oil* (August 1974).
13. *World Oil* (August 1958).
14. *World Oil* (August 1985).
15. *World Oil* (August 1959).
16. Ibid., p. 221.
17. Crawford Young, *Ideology and Development in Africa* (New Haven: Yale University Press, 1982), p.222.
18. Othman, p. 117.
19. *World Oil* (August 1975).
20. *World Oil* (August 1976).
21. Othman, p. 118.

22. Andrade and Olivier, *Angola* (1972), p.74.
23. The first shipment was to Canada; after that, Gulf shipped Cabinda crude to the Northeastern states of the United States, including Massachussetts.
24. *World Oil* (August 1974).
25. John Stockwell, *In Search of Enemies* (New York: W.W. Norton & Co., 1978), p. 204.
26. Makidi-Ku-Ntima, "Class Struggle and the Making of the Revolution in Angola" *Journal of the Insitute for the Study of Labor and Economic Crisis, Contemporary Marxism* 6 (Spring 1983):137.
27. *World Oil* (August 1985).
28. World Bank, *Trends in Developing Economies* (1991), see "Angola."
29. Basil Davidson, "In Angola Now," *West Africa* 3135 (August 8, 1977):133-48.
30. International Monetary Fund, *Surveys of African Economies: Volume 1* (Washington D.C.: IMF, 1968), pp. 57-63.
31. *World Oil* (August 1982).
32. World Bank, *Trends in Developing Countries.*
33. Ibid.
34. Ibid.
35. *World Oil* (August 1960).
36. Young, p. 37.
37. *World Oil* (August 1984).
38. World Bank, op. cit.
39. *World Oil* (August 1988).
40. *World Oil* (August 1981).
41. Othman, p. 135.
42. Sayre Schatz, *Nigerian Capitalism* (Berkeley: University of California Press, 1977), p. 6.
43. Ibid.
44. *World Oil* (August 1978).
45. *World Oil* (August 1982).
46. Alex Newton, *Central Africa: A Travel Survival Kit* (Berkeley: Lonely Planet Publications, 1989), p. 70.
47. Jean-François Bayart, "Cameroun" in Donald O'Brien, John Dunn, and Richard Rathbone, eds., *Contemporary West African States* (Cambridge: Cambridge University Press, 1989), p. 44.
48. Ibid., pp. 39-40.
49. Ibid., p. 45.
50. International Monetary Fund, *Surveys of African*

Economies, Volume 1 (Washington D.C.: IMF, 1968), p. 255.

51. Mark R. Lipschutz and Kent R. Rasmussen, *Dictionary of African Historical Biography: Second Edition* (Berkeley: University of California Press, 1986), p. 284.
52. Newton, p. 141.
53. Daniel Yergin, *The Prize: The Epic Quest for Oil, Money and Power* (New York: Simon and Schuster, 1991), p. 13.
54. Alexander Arbatov, "Oil as a factor in strategic policy and action: past and present," in Arthur Westing, ed., *Global Resources and International Conflict* (1987).

Bibliography

Amin, Samir. *Unequal Development: An Essay on the Social Formation of Peripheral Capitalism*. New York: Monthly Review Press, 1976.

Amnesty International. *Gabon: Deni de justice au cours d'un procés*. Paris: Amnesty International, 1984.

Aubame, Jean-Hilaire. "La Conférence de Brazzaville," in *Afrique Equatoriale Française*. Paris: Encyclopédie Coloniale et Maritime, 1950.

Bach, Daniel. "France's Involvement in Sub-Saharan Africa: A Necessary Condition to Middle Power Status in the International System," in A. Sesay, *Africa and Europe*. London: Croom Helm, 1986.

Ballard, John. "The Development of Political Parties in French Equatorial Africa," Ph.D. dissertation, Fletcher School of Law and Diplomacy, 1964.

Barnes, James. *Gabon: Beyond the Colonial Legacy*. Boulder: Westview Press, 1992.

Bayart, Jean-François. *La politique africaine de François Mitterrand*. Paris: Karthala, 1984.

———. *L'Etat en Afrique: La politique du ventre*. Paris: Fayard, 1989.

Beblawi, Hazem, and Giacomo Luciani, eds. *The Rentier State*. Vol. 2, *Nation, State and Integration in the Arab World*. London: Croom Helm, 1987.

Benda, Julien. *The Treason of the Intellectuals (La Trahison des Clercs)*. New York: W.W. Norton, 1969.

Bierstecher, Thomas J. *Multinationals; the State, and Control of the Nigerian Economy*. Princeton: Princeton University Press, 1987.

Bina, Cyrus. *Economics of the Oil Crisis: Theories of Oil Crisis,*

Oil Rent, and Internationalization of Capital in the Oil Industry. New York: St. Martin's Press, 1985.

Bongo, Omar. *Dialogue et Participation.* Monaco: Paul Bory, 1973.

————. *El Hadj Omar Bongo: Par Lui Même.* Libreville: Editions Multipress, 1983.

————. *Gouverner le Gabon.* Libreville: Editions, 1968.

Bourmaud, Daniel. "France in Africa: African Politics and French Foreign Policy," *Issue: A Journal of Opinion* 23/2 (1995).

Brewer, Anthony. *Marxist Theories of Imperialism: A Critical Survey.* London: Routledge and Kegan Paul, 1980.

Bustin, Edouard. "Une certaine idée de l'Afrique: De Gaulle's vision of Africa between mythology and pragmatism," *Brazzaville +50.* Boston: Groupe de Recherches sur l'Afrique Francophone, 1994.

Cardoso, Fernando Henrique and Enzo Faletto. *Dependency and Development in Latin America.* Berkeley: University of California Press, 1978.

Césaire, Aimé. "Discours sur le colonialisme," *Présence Africaine* (1950).

Chambrier, Hugues Alexandre Barro. *L'Economie du Gabon: Analyse Politique d'Ajustement.* Paris: Economica, 1990.

Chevalier, J.M.. "Theoretical Elements for an Introduction to Petroleum Economics" in *Market, Corporate Behavior, and the State*, eds., A.P. Jacquemin and H.W. de Jong. The Hague: Martinus Nijhoff, 1976.

Coquery-Vidrovitch, Catherine. *Africa: Endurance and Change South of the Sahara.* Berkeley: University of California Press, 1988.

Darlington, Charles F., and Alice B.. *African Betrayal.* New York: David McKay Co., 1968.

Decraene, Philippe. *L'Afrique centrale.* Paris: Centre des Hautes Etudes sur l'Afrique et l'Asie Modernes, 1993.

Delauney, Maurice. *De la casquette à la jaquette.* Paris: Pensée Universelle, 1982.

Dos Santos, Theotino. "The Structure of Dependence," *American Economic Review* 60 (May 1970).

Eboumy, Guy Marcel. "Gabon: From Windfall Surpluses to the Crisis." Master's Thesis, Development Banking, American University, 1989.

Economist Intelligence Unit. *World Outlook '77.* London: EIU, 1977.

Elf-Gabon. *Annual Report 1977.* Paris: 7 rue Nelaton, 1977.

Evans, Peter. *Dependent Development: The Alliance of Multinational, State, and Local Capital in Brazil.* Princeton: Princeton University Press, 1979.

Evans, Peter, Dietrich Rueschemeyer, and Theda Skocpol. *Bringing the State Back In.* Cambridge: Cambridge University Press, 1985.

First, Ruth. "Libya, class and state in an oil economy." in Peter Nore and Terisa Turner, eds., *Oil and Class Struggle.* London: Zed Press, 1980.

Frank, André Gunder. *Dependent Accumulation and Underdevelopment.* New York: Monthly Review Press, 1979.

Frank, Isiah. *Foreign Enterprise in Developing Countries: A Supplementary Paper of the Committee for Economic Development.* Baltimore: Johns Hopkins University Press, 1980.

Frimpong-Ansah, J. *The Vampire State in Africa: The Political Economy of Decline in Ghana.* Trenton: Africa World Press, 1992.

Gaillard, Philippe. *Foccart Parle: Entretiens avec Philippe Gaillard.* Paris: Fayard/Jeune Afrique, 1995.

Gardinier, David. *Gabon.* Oxford: Clio Press, 1992.

Gaulme, François. *Le Gabon et son ombre.* Paris: Karthala, 1988.

Gelb, Alan, and Associates. "Oil Windfalls: Blessing or Curse?" *World Bank Research Publication.* Oxford: Oxford University Press, 1988.

Gendzier, Irene. *Development Against Democracy: Manipulating Political Change in the Third World.* Washington D.C.: Tyrone Press, 1995.

International Institute for Strategic Studies. *The Military Balance.* London: IISS, 1992.

International Labour Office. *Year Book of Labour Statistics.* Geneva: ILO, 1970.

Jackson, Robert, and Carl Rosberg. *Personal Rule in Black Africa.* Berkeley: University of California Press, 1987.

Klapp, Merrie Gilbert. *The Sovereign Entrepreneur: Oil Policies in Advanced and Less Developed Capitalist Countries.* Ithaca: Cornell University Press, 1987.

Leys, Colin. "Capital Accumulation, Class Formation and Dependency — the Significance of the Kenyan Case," in Ralph Miliband and John Saville (eds.) *The Socialist Register.* London: Merlin Press, 1978.

Locke, John. *Second Treatise of Government.* Indianapolis:

Hackett Publishing Company, 1980.

Machiavelli, Niccolo. *The Prince*. New York: Penguin Books, 1993.

Maganga-Moussavou, Pierre-Claver. *Economic Development — Does Aid Help? A Case Study of French Development Assistance to Gabon*. Washington D.C.: African Communications Liason Service, 1983.

Mahdavy, Hossein. "Patterns and Problems of Economic Development in Rentier States: The Case of Iran." in *Studies in the Economic History of the Middle East*, ed., M.A. Cook. Oxford: Oxford University Press, 1970.

Maison Lazard et Compagnie. *Annual Report on the Economy and Finances of the Republic of Gabon*. Paris: Lazard, 1992.

Malthus, Thomas R.. *An Inquiry into the Nature and Progress of Rent*. 1815.

Marx, Karl. *Capital*. New York: Vintage, 1981.

McNamara, Francis Terry. *France in Black Africa*. Washington D.C.: National Defense University Press, 1989.

Médard, Jean-François. "Le Big Man en Afrique: Esquisse d'Analyse du Politicien Entrepreneur," *L'Année Sociologique*, no. 42, 1992.

Ministère de la Coopération, République Française. *Economie et Plan de Développement: République Gabonaise*. Paris: Ministère de la Coopération, 1963.

Ministère de l'Economie et des Finances. *Code Général des Impôts Directs et Indirects*. Libreville: Ministère de l'Economie et des Finances, 1985.

Ministère de l'Economie Nationale, le Commissariat au Plan. *Premier Plan de Développement Economique et Social de la République Gabonaise*. Libreville: Ministère de l'Economie Nationale, 1966.

Ministère de Planification. *Troisième Plan de Développement Economique et Social de la République Gabonaise*. Libreville: Ministère du Plan, 1975.

Ministère de la Planification. *Cinquième Plan de Développement Economique et Social de la République Gabonaise*. Libreville: Ministère de Plan, 1984.

Ministère du Plan, du Développement et des Participations. *Plan Intérimaire de Développement Economique et Social de la République Gabonaise 1980-1982*. Libreville: Ministère du Plan, 1979.

Minogue, Martin, and Judith Malloy. *African Aims and Attitudes: Selected Documents*. Cambridge: Cambridge University

Press, 1979.

Morel, Edmund D.. *The British Case in French Congo: The Story of Great Injustice, Its Causes and Its Lessons*. London: William Heinemann, 1903.

Morse, Edward L.. "After the Fall: The Politics of Oil." *Foreign Affairs* (Spring 1986).

Nafziger, E. Wayne. *Inequality in Africa: Political Elites, Proletariat, Peasants and the Poor*, 2nd ed. Cambridge: Cambridge University Press, 1989.

Naval Intelligence Division. *French Equatorial Africa*. London: NID, 1942.

N'Nah, Nicaolas Metengué. *Economies et Sociétés au Gabon dans la Première Moitié du XIXe Siècle*. Paris: Harmattan, 1979.

Nouaille-Degorce, Brigitte. "Bilan politique de la coopération." *Projet* (May 1962).

Pakenham, Thomas. *The Scramble for Africa: The White Man's Conquest of the Dark Continent from 1876 to 1912*. New York: Random House, 1991.

Patai, Raphael. *The Arab Mind*. New York: Vintage, 1973.

Patterson, Karl David. *The Northern Gabon Coast to 1875*. Oxford: Clarendon Press, 1975.

Péan, Pierre. *Affaires Africaines*. Paris: Fayard, 1983.

————. *L'Argent Noir: Corruption et sous-développement*. Paris: Fayard, 1988.

————. *L'homme de l'ombre: Eléments d'enquête autour de Jacques Foccart, l'homme le plus mysterieux et le plus puissant de la Ve République*. Paris: Fayard, 1990.

Pierce, William Spangard. *Economics of the Energy Industries*. Belmont, CA: Wadsworth, 1986.

Pourtier, Roland. *Le Gabon*, 2 vols. Paris: Harmattan, 1989.

Présidence de la République Gabonaise. *Loi-Programme 1990-1992 des Investissements Publiques et Parapubliques*. Libreville: Présidence, 1989.

Reed, Michael C.. "Gabon: a Neo-Colonial Enclave of Enduring French Interests." *Journal of Modern African Studies* 25, 2 (1987).

Ricardo, David. *The Principles of Economy and Taxation*. London: Everyman's Library, 1821.

Rondot, Jean. *La Compagnie Française des Pétroles: du Franc-Or au Pétrole-Franc*. New York: Arno, 1977.

Sautter, G.. *De l'Atlantique au fleuve Congo: Une Géographie du sous-peuplement*. Paris: Mouton, 1966.

Saint-Paul, Marc Aicardi de. *Gabon: The Development of a*

Nation. London: Routledge, 1989.

Schneider, *The Oil Price Revolution*. Baltimore: Johns Hopkins Press, 1983.

Schweitzer, Albert. *African Notebook*. Indianapolis: Indiana University Press, 1939.

Seers, Dudley. "The Mechanism of an Open Petroleum Economy." *Social and Economic Studies* 13 (1964).

Spero, Joan Edelman. "Dominance-Dependence Relationships: The Case of France and Gabon." Ph.D. diss., Columbia University, 1973.

Thompson, Virginia, and Richard Adloff. *The Emerging States of French Equatorial Africa*. Stanford: Stanford University Press, 1960.

Ungar, Sanford J.. *Africa: The People and Politics of and Emerging Continent*. New York: Simon & Schuster, 1989.

United States Department of Commerce, "Report on Economic and Commercial Climate of Gabon." Washington D.C.: Dept of Commerce, 1991.

Veblen, Thorstein. *The Theory of the Leisure Class*, 1899.

Veyne, Paul. "The Roman Empire." in *From Pagan Rome to Byzantium*. Vol. 1 of *A History of Private Life*, ed. Phillip Ariés. Cambridge: Harvard University Press, 1987.

Warren, Bill. "Imperialism and Capitalist Industrialization," *New Left Review* 81 (September-October 1973).

Weinstein, Brian. *Gabon: Nation-Building on the Ogooué*. Cambridge: MIT University Press, 1966.

World Bank, *Trends in Developing Countries 1990-1991*. New York: World Bank Publications, 1991.

Yergin, Daniel. *The Prize: The Epic Quest for Oil, Money & Power*. New York: Simon & Schuster, 1991.

Young, Crawford. *Ideology and Development in Africa*. New Haven: Yale University Press, 1982.

Index

Sylvoz, Henri, 52
taxation & legitimacy, 34, 36
Tennessee Co. (TENNECO),
    73, 219
Terraroz, Claude, 112
Transgabonese Railroad,
    125, 173–183, 198, 220
Treaty of Vienna, 89
Union Démocratique et
    Sociale Gabonaise (UDSG),
    101, 102, 103, 105, 118
vampirism, 34–35
Warren, Bill, 4, 9
Weinstein, Brian, 81, 96,
    137–138
Wintershall AG, 71, 72
World Energy Development
    Corp. (WED), 72
Yergin, Daniel, 54, 81, 239
Young, Crawford, 201, 227,
    237